Lecture Notes in Mathematics

Edited by A. Dold, Heidelberg and B. Eckmann, Zürich

339

G. Kempf . F. Knudsen
D. Mumford . B. Saint-Donat

Harvard University, Cambridge, MA/USA

Toroidal Embeddings I

Springer-Verlag
Berlin . Heidelberg . New York 1973

AMS Subject Classifications (1970): 14E25, 14D20, 14M99, 20G15

ISBN 3-540-06432-X Springer-Verlag Berlin · Heidelberg · New York
ISBN 0-387-06432-X Springer-Verlag New York · Heidelberg · Berlin

Offsetdruck: Julius Beltz, Hemsbach/Bergstr.

Table of Contents

Introduction

The goal of these notes is to formalize and illustrate the power of a technique which has cropped up independently in the work of at least a dozen people, but which does not seem to have been generally recognized as yet. When teaching algebraic geometry and illustrating simple singularities, varieties, and morphisms, one almost inevitably tends to choose examples of a "monomial" type: i.e., varieties defined by equations

$$X_1^{a_1} \cdots X_r^{a_r} = X_{r+1}^{a_{r+1}} \cdots X_n^{a_n}$$

and morphisms f for which

$$f^*(Y_i) = X_1^{a_{i1}} \cdots X_n^{a_{in}}.$$

Moreover, even when a variety as a whole is quite general, either its singularities, or a certain blow-up may well be defined in local analytic coordinates by monomials. When this happens, problems of algebraic geometry can sometimes be translated into purely combinatorial problems involving the lattice of exponents. This is the technique that we wish to systematize. In particular, the first object of study, considered in Chapter I, is a normal "monomial" variety X; if you wish to express this property without explicit use of coordinates, it turns out to mean that there is an open set $T \subset X$ which is a torus group (an algebraic group isomorphic to $(\mathbb{G}_m)^r$, some $r \geq 1$) such that T acts on X extending its translation action on itself. (We call this a torus embedding) In Chapter II, we go on to consider a variety which is only "locally" of this type. Precisely, we consider $U \subset X$, U open and non-singular in a normal X such that equivalently:

a) $\forall\ x \in X$, the pair $(\hat{\mathcal{O}}_{x,X}, I(X-U)\cdot\hat{\mathcal{O}}_{x,X})$, where $I(X-U)$ = ideal of $X-U$, is isomorphic to $(\hat{\mathcal{O}}_{t,Z}, I(Z-T)\cdot\hat{\mathcal{O}}_{t,Z})$ for some torus embedding $T \subset Z$; and all components of $X-U$ are normal,

or b) $\forall\ x \in X$, $\exists$ a neighborhood U_x of x and an étale map

$$\pi\colon U_x \longrightarrow Z$$

$$\pi^{-1}(T) = U \cap U_x$$

where $T \subset Z$ is a torus embedding.

We call this a toroidal embedding. In both the torus and the toroidal case, we can associate a polyhedral complex Δ to these embeddings and we can translate for instance many birational questions involving blow-ups of X along $X-U$ into questions about Δ. In particular, we apply the machinery to prove a theorem about the existence of "semi-stable reductions" in all dimensions - cf. Introduction to Ch. II for a complete statement. For this Theorem, the combinatorial problem on the Δ level is solved in Ch. III. Finally, in Chapter IV we connect our Δ's with some graphs of Orlik and Wagreich and with Tits' buildings. We intend to write a sequel to these notes explaining the application of the same technique to constructing explicit resolutions of the projective varieties $\overline{D/\Gamma}$ (where D is a bounded, symmetric domain, Γ is an arithmetic group, and $\overline{}$ refers to the Borel-Baily "minimal" compactification). We would also like to study semi-stable reduction over a higher dimensional base: viz., given any dominating morphism $f: X \longrightarrow Y$, replacing Y by any Y' generically finite and proper over Y and X by a blow-up of the component of $X \times_Y Y'$ dominating Y', simplify all the fibres $f': X' \longrightarrow Y'$ as much as possible while requiring that X',Y' are non-singular and f is flat.

As stated above, the ideas here have many sources. In particular, I have been influenced for quite a while by the unpublished parts of Hironaka's Ph.D thesis on "The characteristic cone attached to a birational morphism", and a (apparently unpublished) preprint by John Nash entitled "The arc structure of singularities". More specifically, the motivating problem for me has been how to blow up the cusps of Satake's compactification of the abelian variety moduli space $\mathcal{A}_n = D_n/\Gamma_n$ (where $D_n = U(n)\backslash Sp(2n,\mathbb{R})$, $\Gamma_n = Sp(2n,\mathbb{Z})$). We should refer here to the work of Tate and Morikawa (Theta functions and abelian varieties over valuation rings, Nagoya Math. J., 20 (1962)) on p-adic uniformizations; joint work of mine with Alan Mayer (see Further comments on boundary points, in Lecture Notes from the AMS Summer Institute on Algebraic Geometry, Woods Hole, 1964) and my own version published in "An analytic construction of degenerating abelian varieties over complete rings, Comp. Math. 24 (1972); especially to Igusa's paper "A desingularization problem in the theory of Siegel modular functions, Math. Annalen, 168 (1967); to the work of Hirzebruch on desingularization for the Hilbert modular group (cf. Seminaire Bourbaki, exp. 396 (1971)); and to recent work by Satake announced in his AMS lecture "On the arithmetic of Tube domains", to appear in the Bulletin of the AMS. On the other hand, the concept of a torus embedding appears for the first time in the article of Demazure, Sous-groups algebriques de rang maximum du groupe de Cremona, Annales Sci. Ecole Norm. Sup., 3 (1970). Independently, Hochster considered very similar constructions in "Rings of invariants of tori, Cohen-Macaulay rings generated by monomials and polytopes, Annals of Math. 96 (1972)*. Related ideas are also found in Bergman, The logarithmic limit-set of an algebraic variety, Trans. Am. Math. Soc. 157 (1971). A crucial result showing that all torus embeddings are covered by affine ones (Ch. I, §2, Th. 5) is due to H. Sumihiro, Equivariant Completion (to appear). As for the problem of semi-stable reduction itself, the curve case has been known for some time: the best proof is in Artin and Winters, Degenerate fibres and reduction of curves, (Topology, 10 (1971) p.373). The real breakthrough in the higher-dimensional case was an elegant combinatorial construction due to Alan Waterman, described in Ch. III, §3. I talked about this and char. 0 semi-stable reduction in all dimensions in Houston last year: but the next month I found a big gap and the final proof which we give here is due jointly to F. Knudsen, A. Waterman and myself. The problem of semi-stable reduction in char. p and dimension > 1 is open and probably very difficult.

David Mumford

*) After this was written, I received a paper by K.Miyake and T.Oda entitled Almost homogeneous algebraic varieties under algebraic torus action also on this topic.

Note: All references will be given in the text, except: EGA=Elements de Geomtrie Algebrique, A.Grothendieck; Z-S = Commutative Algebra, O.Zariski and P.Samuel; G.I.T. = Geometric Invariant Theory, D.Mumford; I= Chapter 1 of these notes.

Chapter I

Equivariant Embeddings of Tori

k is an algebraically closed field; a variety is an integral scheme of finite type, separated, over k.

As usual $\mathbb{G}_m = GL(1)$ denotes the algebraic group whose points over k are the non zero elements of k.

An algebraic torus is an algebraic group T isomorphic to $\mathbb{G}_m^n$ where n is an integer ≥ 1: the purpose of this chapter is to study the varieties X containing T as an affine open subset, such that the translations on T extend to an action of T on X; such an X will be called an equivariant embedding of T.

§1. Affine embeddings

For the basic facts about tori, we refer the reader to Borel's Linear Algebraic Groups (Benjamin) and we just fix here our notations.

Let T be an n-dimensional torus: by definition there exist indeterminants $X_1,\cdots,X_n$ such that $\Gamma(\mathcal{O}_T)$ is isomorphic to $k[X_1,X_1^{-1},X_2,X_2^{-1},\cdots,X_n,X_n^{-1}]$.

Let $M = \mathrm{Hom}_{\text{alg. groups}}(T,\mathbb{G}_m)$ the group of characters of T: if $r \in M$, we denote by $\mathcal{X}^r$ the corresponding element of $\Gamma(\mathcal{O}_T^*)$.

Let $N = \mathrm{Hom}_{\text{alg. groups}}(\mathbb{G}_m,T)$ the group of 1-parameter subgroups of T (we abbreviate this to 1-PS of T): if $a \in N$, we denote by $\lambda_a\colon \mathbb{G}_m \longrightarrow T$ the corresponding homomorphism.

M and N are free abelian groups of rank n, related by a non singular canonical pairing:

$$M \times N \longrightarrow \mathbb{Z}$$
$$(r,a) \longmapsto \langle r,a \rangle$$

where $\langle r,a \rangle$ is defined by the condition

$$\chi^r(\lambda_a(t)) = t^{\langle r,a \rangle}, \qquad t \in k^* .$$

With these notations, the χ^r, $r \in M$, form a basis of $\Gamma(\mathcal{O}_T)$ as a k-vector space, and

$$\Gamma(\mathcal{O}_T) = k[\cdots, \chi^r, \cdots]_{r \in M}$$

is canonically identified to the <u>algebra of the group</u> M, also denoted k[M].

Now, a choice of a system of coordinates

$$\Gamma(\mathcal{O}_T) \cong k[X_1,X_1^{-1},X_2,X_2^{-1},\cdots,X_n,X_n^{-1}]$$

defines an isomorphism $\mathbb{Z}^n \cong M$ under which $r \in M$ corresponds to $(r_1,\cdots,r_n) \in \mathbb{Z}^n$ and χ^r corresponds to $X_1^{r_1}X_2^{r_2}\cdots X_n^{r_n}$.

Recall that the data of an action of T on an affine variety X = Spec B is equivalent to the data of a gradation of type M:

$$B = \bigoplus_{r \in M} B_r$$

where B_r is the set of functions $f \in B$ such that

$$f(\alpha x) = \chi^r(\alpha) f(x)$$

for all R-valued points α and x of T and X respectively, where R is any k-algebra. Moreover, equivariant homomorphisms correspond to homomorphisms of graded algebras.

Definition 1: An equivariant affine embedding of T (resp. an equivariant embedding of T) is an affine variety (resp. a variety) containing T as an open subset, equipped with an action of T:

$$T \times X \longrightarrow X$$

extending the action $T \times T \longrightarrow T$ given by the translations in T. A morphism between two equivariant embeddings X and X' of T is a map f such that the following diagram:

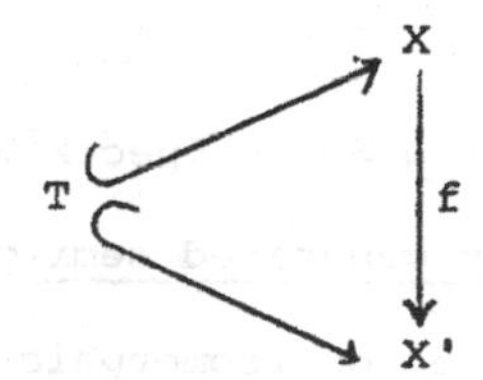

commutes.

Let S be a semi-group $\subset$ M, which is finitely generated (as a semi-group); we denote by k[S] the subvector space of $\Gamma(\mathcal{O}_T)$ generated by the $\mathfrak{X}^r$, $r \in S$: k[S] is an algebra of finite type over k which can be viewed as the algebra of the semi-group S.

Moreover, let us note the following fact:

The quotient field of k[S] is equal to the quotient field of $\Gamma(\mathcal{O}_T)$ if and only if S generates M as a group.

To see this, let us denote by M' the subgroup of M generated by S: we can find a system of coordinates (i.e., an isomorphism $\mathbb{Z}^n \xrightarrow{\sim} M$) and positive integers $d_1,\cdots,d_k$ $(1 \leq k \leq n)$, such that

$$(d_1,0,\cdots,0),(0,d_2,0,\cdots,0),\cdots,(0,\cdots,0,d_k,0,\cdots,0)$$

is a basis for M'. Then the quotient field of $k[S]$ is equal to $k(X_1^{d_1},\cdots,X_k^{d_k})$ which is equal to $k(X_1,X_2,\cdots,X_n)$ if and only if $k = n$ and $d_i = 1$ for all i. q.e.d.

If S generates M as a group, then the inclusion $k[S] \subset \Gamma(\mathcal{O}_T)$ induces an equivariant embedding $T \subset \operatorname{Spec} k[S]$ and it follows from the preceding remarks that we get in this way all the equivariant affine embeddings, so we proved:

Proposition 1: The correspondence $S \longmapsto \operatorname{Spec} k[S]$ defines a bijection between the set of finitely generated semi-groups $S \subset M$ which generate M as a group and the set of isomorphic classes of equivariant affine embeddings of T. Moreover, the morphisms of equivariant affine embeddings correspond (in a contravariant way) to the inclusions between semi-groups $\subset M$.

Definition 2. Let $S \subset M$ be a semi-group: we say that S is saturated if the condition $nr \in S$, where n is an integer ≥ 1 and r an element of M, implies $r \in S$.

With this definition, we have

Lemma 1: *Let $T \subset X$ be an equivariant affine embedding of T corresponding to a semi-group S in M. Then X is normal if and only if S is saturated.*

Proof: (1) Let us assume that $k[S]$ is normal: if $r \in M$ and n is an integer ≥ 1 such that $nr \in S$, then $\mathfrak{X}^r$ satisfies the equation $X^n - \mathfrak{X}^{nr} = 0$ of integral dependence over $k[S]$, so $r \in S$ and S is saturated.

(2) The condition is sufficient:

Let R' be the normalization of $k[S]$: we have the inclusions $k[S] \subset R' \subset \Gamma(\mathcal{O}_T)$ and it follows from [ZS, vol. II, p. 158] that an element $f \in \Gamma(\mathcal{O}_T)$ belongs to R' if and only if all its homogeneous components belong to R'. Therefore if we assume that $k[S]$ is not normal, there exists an element $r \in M$ such that $\mathfrak{X}^r$ is integral over $k[S]$, but $r \notin S$:

Let

$$X^n + a_{n-1}X^{n-1} + a_{n-2}X^{n-2} + \dots + a_0 = 0$$

be an equation of integral dependence for $\mathfrak{X}^r$, where $a_i \in k[S]$ for all i. Writing each a_i as a sum of characters, we may throw out all those characters for which the degree doesn't add up, i.e., make

$$nr = r(n-i) + \deg a_i, \qquad 1 \leq i \leq n .$$

Now if $a_i \neq 0$, then $\deg a_i \in S$, so $r \cdot i \in S$ and $r \in S$ by saturation.

q.e.d.

Combining Proposition 1 and Lemma 1, we get:

<u>Theorem 1</u>: <u>The correspondence</u> $S \longmapsto \operatorname{Spec} k[S]$ <u>defines a bijection between the set of finitely generated semi-groups</u> $S \subset M$ <u>which generate</u> M <u>as a group and which are saturated and the set of equivariant normal affine embeddings of</u> T.

We are now going to interpret the semi-groups which occur in Theorem 1 in terms of simplicial objects sitting canonically in $M_{\mathbb{R}} = M \otimes_{\mathbb{Z}} \mathbb{R}$ or $N_{\mathbb{R}} = N \otimes_{\mathbb{Z}} \mathbb{R}$.

Let us note first that for a subset σ of $N_{\mathbb{R}}$ the two following conditions are equivalent:

(i) there exist a finite number of linear functionals ℓ_i, $i = 1,\cdots,N$, defined over $\mathbb{Q}$, such that

$$\sigma = \left\{ x \,\middle|\, \ell_i(x) \geq 0 \quad \text{for all } i \right\}.$$

(ii) there exist a finite number of vectors $x_i \in N_{\mathbb{R}}$, $i = 1,\cdots,N$, defined over $\mathbb{Q}$, such that

$$\sigma = \left\{ \sum_{i=1}^{N} \lambda_i x_i \,\middle|\, \lambda_i \geq 0, \text{ all } i \right\}.$$

Such a set will be called a <u>convex rational polyhedral cone</u>.

Now let $\check{\sigma} \subset M_{\mathbb{R}}$ be the set of r such that $\langle r,a\rangle \geq 0$ for all $a \in \sigma$; then $\check{\sigma}$ is also a convex rational polyhedral cone and $\check{\check{\sigma}} = \sigma$: if $\sigma = \left\{ x \,\middle|\, \ell_i(x) \geq 0,\ i = 1,\cdots,N \right\}$ then $\check{\sigma}$ is the set of $\sum_{i=1}^{N} \lambda_i \ell_i$ where $\lambda_i \geq 0$ for all i. Moreover σ doesn't contain any linear subspace if and only if $\check{\sigma}$ is not contained in a hyperplane.

Let $\sigma = \{x \mid \ell_i(x) \geq 0,\ i = 1,\cdots,N\}$ be a convex rational polyhedral cone: then a face of σ is a subset σ' of σ of the form $\sigma \cap \{x \mid \ell(x) = 0\}$ where ℓ is an element of $\check{\sigma}$: in other words, there exists a subset I of $[1,N]$ such that $\sigma' = \sigma \cap \{x \mid \ell_i(x) = 0$ for $i \in I\}$. The interior of σ' is the set of elements x of σ' such that $\ell_i(x) > 0$ for all $i \notin I$.

Lemma 2: The correspondence $\sigma \longmapsto \check{\sigma} \cap M$ defines a bijection between the set of convex rational polyhedral cones in $N_{\mathbb{R}}$ which do not contain any linear subspace and the set of finitely generated semi-groups $S \subset M$, which generate M as a group.

Proof: If σ is a convex rational polyhedral cone which do not contain any linear subspace, then $\check{\sigma}$ is not contained in a hyperplane and $\check{\sigma} \cap M$ is a saturated semi-group which generates M as a group. The fact that $\check{\sigma} \cap M$ is finitely generated is expressed in the following statement, the proof of which is left to the reader.

Gordan's lemma*: Given a finite set of homogeneous linear integral inequalities, the semi-group of integral solutions is finitely generated.

Conversely let S be a saturated semi-group which admits a finite set of generators x_i, $i = 1,\cdots,N$. Let

*Actually Gordan proved the essentially equivalent fact that, given a finite set of homogeneous linear equations, then the semi-group of non-negative integral solutions is finitely generated—cf. Vorlesungen über Invariantentheorie, vol. 1., p. 199 (Verlag B.G. Teubner, Leipzig, 1885).

$\check{\sigma} = \left\{ \sum_{i=1}^{N} \lambda_i x_i \,\middle|\, \lambda_i \in \mathbb{R},\ \lambda_i \geq 0 \right\}$ be the convex rational polyhedral cone spanned by the x_i. If $r \in \check{\sigma} \cap M$, then r can be written $\sum_{i=1}^{N} q_i x_i$ with $q_i \in \mathbb{Q}$, $q_i \geq 0$: so there exists an integer n such that $nr \in S$ and $r \in S$ by saturation, hence $S = \check{\sigma} \cap M$. Moreover $\check{\sigma}$ is not contained in a hyperplane if S generates M as a group.

Now we can state the main result of this section:

Theorem 1': *The correspondence* $\sigma \longmapsto \operatorname{Spec} k[\check{\sigma} \cap M] = X_\sigma$ *defines a bijection between the set of rational convex polyhedral cones in* $N_{\mathbb{R}}$ *which do not contain any linear subspace and the set of equivariant affine normal embeddings of* T. *Moreover, if* $a \in N$, $a \in \sigma$ *if and only if* $\lim_{t \to 0} \lambda_a(t)$ *exists in* X_σ.

The first part of Theorem 1' follows immediately from Theorem 1 and Lemma 2.

Let λ_a be a 1-PS of T: then $\lim_{t \to 0} \lambda_a(t)$ exists in X_σ if and only if $\lim_{t \to 0} \mathfrak{X}^r(\lambda_a(t))$ exists in $\mathbb{A}^1_k$ for all $r \in \check{\sigma} \cap M$. But $\mathfrak{X}^r(\lambda_a(t)) = t^{\langle r,a \rangle}$, therefore $\lim_{t \to 0} \mathfrak{X}^r(\lambda_a(t))$ exists if and only if $\langle r,a \rangle \geq 0$, and the last part of Theorem 1' follows from our definitions.

Remark. This simple characterization of σ in terms of the variety X_σ is the reason why we looked at cones sitting in $N_{\mathbb{R}}$ rather than $M_{\mathbb{R}}$.

Notation: If $a \in N \cap \sigma$, the closed point $\lim_{t \to 0} \lambda_a(t)$ in X_σ will be denoted $\lambda_a(0)$.

For the remainder of this section, we interpret the structure of the X_σ's (as T-spaces) and the map between them in terms of convex polyhedral cones in $N_{\mathbb{R}}$.

Theorem 2. Let σ be a convex rational polyhedral cone and X_σ the affine equivariant embedding of T associated to σ by Theorem 1', then:

(a) Let a_1 and a_2 be two elements of $\sigma \cap N$, then $\lambda_{a_1}(0) = \lambda_{a_2}(0)$ if and only if a_1 and a_2 lie in the interior of the same face of σ.

(b) In every orbit $\mathcal{O}$ in X_σ (under the action of T), there is a unique point of the type $\lambda_a(0)$, for some $a \in N \cap \sigma$.

(c) There is a bijection $\sigma' \longmapsto \mathcal{O}^{\sigma'}$ between the set of faces of σ and the T-orbits in X_σ.

(d) $\sigma_1 \subset \sigma_2$ if and only if $\overline{\mathcal{O}^{\sigma_1}} \supset \mathcal{O}^{\sigma_2}$.

(e) $\dim \sigma + \dim \mathcal{O}^\sigma = n$.

Proof of Theorem 2.

(a) Let $\{r_1,\cdots,r_N\}$ be a system of generators of S, so that $\{\mathfrak{X}^{r_1},\ldots,\mathfrak{X}^{r_N}\}$ can be chosen as a system of coordinates for X_σ. Let $a \in \sigma \cap N$, then $\mathfrak{X}^{r_i}(\lambda_a(t)) = t^{\langle r_i,a\rangle}$ and

$$(*) \qquad \mathfrak{X}^{r_i}(\lambda_a(0)) = \begin{cases} 1 & \text{if } \langle r_i,a\rangle = 0 \\ 0 & \text{if } \langle r_i,a\rangle > 0 . \end{cases}$$

In other words, $\lambda_a(0)$ is determined by the subset $I \subset \{1,\cdots,N\}$ defined by the condition: $\langle r_i,a\rangle > 0$ for all $i \in I$.

(b) Let $\mathcal{O}$ be a T-orbit of X_σ and x a closed point of $\mathcal{O}$. Let $A = k[[t]]$. We can find* an A-valued point τ of X_σ

---

*This can be proved by (1) via any n algebraically independent formal power series $f_i(t)$ with $f_i(0) = 0$, defining

$$\tau_0: \operatorname{Spec} A \longrightarrow \mathbb{A}^n$$
$$\tau_0 \text{ (closed pt.)} = 0$$
$$\tau_0 \text{ (generic pt.)} = \text{generic pt.},$$

and (2) representing X_σ as finite over $\mathbb{A}^n$ by Noether normalization, lifting τ_0 to a morphism τ of a suitable finite covering Spec A' of Spec A to X_σ and noting that the integral closure of A' is $\cong k[[t']]$ by Cohen's structure theorems.

$$\tau : \operatorname{Spec} A \longrightarrow X_\sigma$$

such that the closed point (resp. the generic point) is mapped on x, (resp. on the generic point of X_σ).

If $X_1, X_2, \cdots, X_n$ is a coordinate system for T, τ defines an extension

$$k(X_1, \cdots, X_n) \longrightarrow k((t)).$$

Write $X_i = t^{a_i} . u_i(t)$, where $a_i \in \mathbb{Z}$, $u_i(t) \in k[[t]]$, $u_i(0) \neq 0$ for all $i = 1, \cdots, n$: $(u_1(t), \cdots, u_n(t))$ defines an A-valued point μ of T such that

$$(\mu^{-1} . \tau)(0) = (\mu(0))^{-1} . \tau(0) \in \mathcal{O}.$$

But this point is the limit of the 1-P.S

$$t \rightsquigarrow (t^{a_1}, t^{a_2}, \ldots, t^{a_n}) .$$

So $\mathcal{O}$ contains a point of the form $\lambda_a(0)$, where $a \in N \cap \sigma$.

Let a_1, a_2 be two elements of $N \cap \sigma$ such that $\lambda_{a_1}(0) = \alpha \lambda_{a_2}(0)$ where $\alpha \in T(k)$: then for $i = 1, \cdots, N$

$$\mathfrak{X}^{r_i}(\lambda_{a_1}(0)) = \mathfrak{X}^{r_i}(\alpha) . \mathfrak{X}^{r_i}(\lambda_{a_2}(0)).$$

But then the proof of (a) shows that $\lambda_{a_1}(0) = \lambda_{a_2}(0)$.

The combination of (a) and (b) shows (c).

(d) Let $\mathcal{O}^{\sigma_1}$ be defined by $\mathfrak{X}^{r_i} \equiv 0$, $i \in I_1 \subset \{1, \cdots, N\}$ and $\lambda_{a_1}(0) \in \mathcal{O}^{\sigma_1}$,

let $\mathcal{O}^{\sigma_2}$ be defined by $\mathfrak{X}^{r_i} \equiv 0$, $i \in I_2 \subset \{1, \cdots, N\}$ and $\lambda_{a_2}(0) \in \mathcal{O}^{\sigma_2}$.

Then we have

$$\begin{array}{ccc} \mathcal{O}^{\sigma_2} \subset \overline{\mathcal{O}^{\sigma_1}} & \Longrightarrow & I_1 \subset I_2 \\ \Uparrow & & \Downarrow \\ & \lim_{t \to 0} \lambda_{a_2}(t)\lambda_{a_1}(0) = \lambda_{a_2}(0) \ . & \end{array}$$

On the other hand, $I_1 \subset I_2$ if and only if $\sigma_1 \subset \sigma_2$.

(e) Let $a \in \sigma \cap N$ and μ be a 1-PS of T. Then $\lambda_a(0)$ is fixed under the action of $\mu(t)$ for all $t \in k^*$ if and only if

$$\forall\ i = 1,\cdots,N \quad , \quad \mathcal{X}^{r_i}(\lambda_a(0)) = 0 \quad \text{or} \quad \mathcal{X}^{r_i}\cdot\mu \equiv 1;$$

in other words

$$\forall\ i = 1,\cdots,N \quad , \quad \langle r_i,a\rangle > 0 \quad \text{or} \quad \langle r_i,\mu\rangle = 0.$$

But this last condition is equivalent to the fact that μ belongs to the linear subspace spanned by the face containing a, hence

$$\dim(\text{face containing a}) = \dim(\text{isotropy group of } \lambda_a(0)) \ .$$

Remark. By Theorem 2, the vertex of σ corresponds to the unique open orbit T in X_σ and the interior of σ to the lowest dimensional orbit, which is also the unique closed orbit.

Theorem 3. Let X_{σ_1} and X_{σ_2} be two equivariant affine embeddings on T. Then there exists a map $X_{\sigma_1} \xrightarrow{f} X_{\sigma_2}$ such that the following diagram commutes:

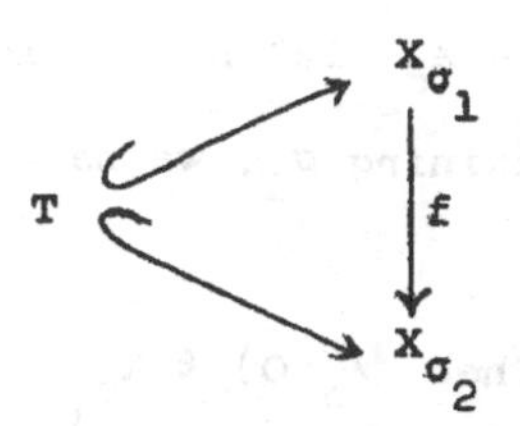

<u>if and only if</u> $\sigma_1 \subseteq \sigma_2$. <u>Moreover</u> f <u>is unique</u>, <u>equivariant</u>, <u>and</u> f <u>is an open immersion if and only if</u> σ_1 <u>is a face of</u> σ_2.

<u>Proof</u>. (1) Assume that f exists: if $a \in \sigma_1 \cap N$ then $\lambda_a(0)$ exists in X_{σ_1}, therefore $\lambda_a(0)$ exists in X_{σ_2} and $a \in \sigma_2 \cap N$ by Theorem 1'.

(2) Conversely if $\sigma_1 \subseteq \sigma_2$ then $\check{\sigma}_2 \subset \check{\sigma}_1$ and this inclusion induces an inclusion $k[\check{\sigma}_2 \cap M] \subset k[\check{\sigma}_1 \cap M]$ which defines a map $X_{\sigma_1} \longrightarrow X_{\sigma_2}$.

(3) Assume that σ_1 is a face of σ_2: then by definition there exists an $\alpha \in M$ such that $\alpha \geq 0$ on σ_2 and such that $\sigma_1 = \{x \mid \alpha(x) = 0\} \cap \sigma_2$.

Now since $\alpha(x) > 0$ for $x \in \sigma_2 - \sigma_1$, for every $\beta \in M$ such that $\beta \geq 0$ on σ_1, there exists an integer $n \geq 1$ such that $\beta + n\alpha \geq 0$ on σ_2, hence:

$$\Gamma(\mathcal{O}_{X_{\sigma_1}}) = k[\cdots, \mathfrak{X}^{\beta}, \cdots]_{\substack{\beta \geq 0 \\ \text{on } \sigma_1}} = k[\cdots, \mathfrak{X}^{\gamma}, \cdots]_{\substack{\gamma \geq 0 \\ \text{on } \sigma_2}} \left[\frac{1}{\mathfrak{X}^{\alpha}}\right] = \Gamma(\mathcal{O}_{X_{\sigma_2}})\left[\frac{1}{\mathfrak{X}^{\alpha}}\right]$$

and $X_{\sigma_1} \longrightarrow X_{\sigma_2}$ is an open immersion.

(4) Assume that $X_{\sigma_1} \longrightarrow X_{\sigma_2}$ is an open immersion: let τ be the smallest face of σ_2 containing σ_1, we have then an open immersion $X_{\sigma_1} \subset X_\tau$.

Let $a \in \sigma_1 \cap \mathrm{Int}(\tau)$. Then $\lambda_a(0) \in X_{\sigma_1} \cap X_\tau$.

Now if $\mathcal{O}'$ is any orbit of X_τ under T, we have

$$\mathcal{O}(\lambda_a(0)) \subset \overline{\mathcal{O}'} \qquad \text{by Theorem 2 , d).}$$

Therefore X_{σ_1} must meet $\mathcal{O}'$ so $X_{\sigma_1} = X_\tau$ and $\sigma_1 = \tau$.

__Theorem 4__. X_σ __is non singular if and only if__ $\sigma \cap N$ __can be generated by a subset of a basis of__ N __over__ $\mathbb{Z}$.

__Proof.__ (1) Let $a_1, \cdots, a_n$ be a basis of N and for some r, $1 \leq r \leq n$, let

$$\sigma = \left\{ \sum_{i=1}^{r} \lambda_i a_i \;\middle|\; \lambda_i \geq 0 \right\} .$$

Then $X_\sigma \cong \mathrm{Spec}\, k\left[X_1, \cdots, X_r, X_{r+1}, X_{r+1}^{-1}, \cdots, X_n, X_n^{-1}\right] \simeq \mathbb{G}_a^r \times \mathbb{G}_m^{n-r}$ is non singular.

(2) Let us assume that X_σ is non singular.

Let $N'_{\mathbb{R}}$ the smallest linear subspace of $N_{\mathbb{R}}$ containing σ, and let $N' = N'_{\mathbb{R}} \cap N$.

There exists a decomposition $N = N' \oplus N''$ (since N/N' is torsion free) which induces a decomposition $M = M' \oplus M''$, $T = T' \times T''$ and, as one sees immediately, $X_\sigma \cong X'_\sigma \times T''$ where X'_σ is the equivariant compactification of T' associated to $\sigma \subset N'_{\mathbb{R}}$. Hence X'_σ is nonsingular, and we are reduced to the case where σ spans $N_{\mathbb{R}}$, or in other words, to the case where X_σ contains a fixed point O (by the Remark following Theorem 2).

It is clear that the ideal I(O) of O in $R = \Gamma(\mathcal{O}_{X_\sigma})$ is generated by the $\mathfrak{X}^\alpha$ where $\alpha \neq 0$ and $\alpha \in \check{\sigma} \cap M = S$. In other words

$$R = k.\mathfrak{X}^0 \oplus I(O).$$

$I(O)R_{I(O)}$ is generated by n elements, therefore there exist $\mathfrak{X}^{\alpha_1},\ldots,\mathfrak{X}^{\alpha_n}$ such that for all $\beta \in S$

$$\mathfrak{X}^\beta = \sum_{i=1}^{n} a_i \mathfrak{X}^{\alpha_i} \qquad \text{where} \quad a_i \in R_{I(O)}$$

or

$$u.\mathfrak{X}^\beta = \sum_{i=1}^{n} b_i \mathfrak{X}^{\alpha_i} \qquad \text{where} \quad u \in R - I(O) \text{ and } b_i \in R.$$

By taking the homogeneous part of "degree β" in the both sides of this last relation, we get (if $\beta \neq 0$).

$$\beta = \gamma + \alpha_i \qquad \text{where} \quad i \in [1,n] \quad \text{and} \quad \gamma \in S.$$

This shows that S is generated as a semi-group by the α_i's, hence the α_i's form a basis of N over $\mathbb{Z}$. Q.E.D.

One final remark on this situation: for every face τ of σ, there is a natural T-equivariant retraction P_τ of X_σ onto the closure of the orbit corresponding to τ:

$$X_\sigma \underset{\hookleftarrow}{\overset{P_\tau}{\longrightarrow}} \overline{\mathcal{O}^\tau}$$

In fact, if $R = \Gamma(\mathcal{O}_{X_\sigma})$, then

$$R = \bigoplus_{\substack{r\geq 0 \\ \text{on } \sigma}} k\cdot\mathfrak{X}^r = \underbrace{\left[\bigoplus_{\substack{r\geq 0 \text{ on } \sigma \\ r\equiv 0 \text{ on } \tau}} k\cdot\mathfrak{X}^r\right]}_{R'} \oplus \underbrace{\left[\bigoplus_{\substack{r\geq 0 \text{ on } \sigma \\ r>0 \text{ on Int } \tau}} k\cdot\mathfrak{X}^r\right]}_{I}$$

Since I is the ideal of $\overline{\mathcal{O}^\tau}$, this shows that there is a T-invariant subring of R mapping isomorphically to R/I. Thus

$$\text{Spec } R' \cong \text{Spec } R/I = \overline{\mathcal{O}^\tau}$$

and the inclusion of R' in R defines the retraction P_τ. It follows easily that the open subset $X_\tau \subset X_\sigma$ is just $P_\tau^{-1}(\mathcal{O}^\tau)$. Moreover R' is integrally closed in its quotient field by Lemma 1 above, hence :

Proposition 2. $\forall$ orbits $\mathcal{O} \subset X_\sigma$, $\overline{\mathcal{O}}$ is a normal variety.

Let us illustrate what has been done so far by looking at the case n = 2:

Let σ be a convex rational polyhedral cone in $N_{\mathbb{R}} \simeq \mathbb{R}^2$ which does not contain any linear subspace and such that Int $\sigma \neq \emptyset$. Then we can find a basis (e_1,e_2) of N such that:

$$\sigma = \left\{\lambda e_1 + \mu(ae_1+be_2) \,\middle|\, \lambda,\mu \in \mathbb{R}^+\right\}$$

where a and b are integers such that $b > 0$ and $(a,b) = 1$ (this last condition just means that ae_1+be_2 is primitive in N).

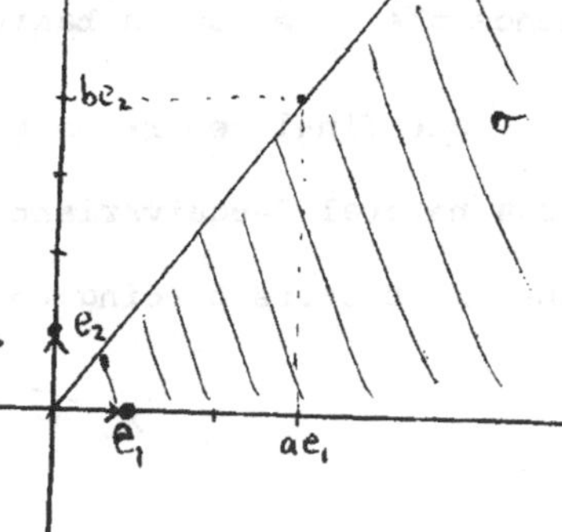

By Theorem 4, X_σ is non-singular if and only if $b = \det(e_1, ae_1+be_2) = 1$: in this case X_σ is isomorphic to $\mathbb{A}^2_k$. Otherwise, when $b > 1$, X_σ has an isolated singular point which constitutes the unique closed orbit. We are going to show that <u>X_σ appears naturally as the quotient of $\mathbb{A}^2_k$ by a finite cyclic group acting diagonally</u>.

Let N^* be the sublattice of N generated by e_1 and ae_1+be_2: $N^* = \left\{ \alpha e_1+\beta e_2 \mid \beta \in b.\mathbb{Z} \right\}$ and $N/N^* \xrightarrow{\sim} \mathbb{Z}/b\mathbb{Z}$.

Let M^* be the set of $x \in M_{\mathbb{R}}$ such that $\langle n,a\rangle \in \mathbb{Z}$ for all $a \in N^*$: $M^* = \left\{ \lambda e_1' + \mu e_2' \mid \mu \in 1/b.\mathbb{Z} \right\}$, where $\{e_1', e_2'\}$ is the dual basis of $\{e_1, e_2\}$, and $\left\{ e_1' - a/b e_2', 1/b e_2' \right\}$ is a basis of M^*.

Let us assume now that $\mathrm{char}(k) \nmid b$ and let ξ be a primitive b^{th} root of 1. Then we can construct a symmetric pairing:

$$e: \quad M^*/M \times N/N^* \longrightarrow \mu_b$$

by setting $e(\bar{m}, \bar{n}) = \xi^{b\langle m,n\rangle}$, where $\bar{m}$ (resp. $\bar{n}$) represents the class modulo M (resp. N^*) of an element m of M^* (resp. n of N): this pairing is non-singular by construction.

Let $\check{\sigma}$ be the dual of σ in $M_{\mathbb{R}}$: $\check{\sigma} = (be_1' - ae_2', e_2')$. The inclusion $\check{\sigma} \cap M \subset \check{\sigma} \cap M^*$ induces a map

$$\mathbb{A}^2_k \xrightarrow{\sim} \operatorname{Spec} k[\check{\sigma} \cap M^*] \longrightarrow \operatorname{Spec} k[\check{\sigma} \cap M] = X_\sigma.$$

Moreover, let $g \in N/N^*$ and $r \in M^*$: by setting $g.\mathfrak{X}^r = e(\bar{r},g)\mathfrak{X}^r$, we define an action of N/N^* on Spec $k[\check{\sigma}\cap M^*]$ so that X_σ appears as the quotient of this action, as it follows from the definitions and the fact that e is non-singular.

To be more explicit, let (x,y) be the system of coordinates for $\mathbb{A}_k^2 \simeq \text{Spec } k[\check{\sigma}\cap M^*]$ corresponding to the basis $\{e_1'-a/be_2', 1/be_2'\}$ of M^*; using the isomorphism $N/N^* \xrightarrow{\sim} \mu_b$ which sends $\bar{e}_2$ to ξ, the preceding action can be written:

$$\xi\cdot(x,y) = (\xi^{-a}x,\ \xi y)\ .$$

If, for example, $a = 1$:

then

$$k[X,Y]^{\mu_b} = k[XY,\ X^b,\ Y^b] \simeq k[u,v,w]/(vu-w^b).$$

We get a rational double point of type A_{b-1}.

If $a = -1$, then

$$k[X,Y]^{\mu_b} = k[X^b, X^{b-1}Y, X^{b-2}Y^2, \cdots, Y^b]$$

which is the affine ring of the cone in $\mathbb{A}^b$ over a rational normal curve of degree b-1 in $\mathbb{P}^{b-1}$.

<u>Remarks</u>. 1) Using the automorphisms of N, whose matrix with respect to (e_1,e_2) is given by $\begin{pmatrix}1 & n\\ 0 & 1\end{pmatrix}$, $n \in \mathbb{Z}$, we are reduced to consider the case $0 < a < b$. These singularities have been studied in detail by E. Brieskorn in [ ].

2) In the n-dimensional case, the method used above to describe X_σ shows that whenever σ is a simplicial cone, i.e., generated by a basis of $N_{\mathbb{R}}$ so that $\sigma = \left\{\sum_{i=1}^{n} \lambda_i e_i \,\middle|\, \lambda_i \geq 0\right\}$, e_i primitive vectors in N generating a subgroup N* of N of finite index k, then

$$X_\sigma \cong \mathbb{A}^n/\mu$$

where $\mu \cong \prod \mu_{k_i}$ and $\prod k_i = k$ and μ acts diagonally on $\mathbb{A}^n$ via n characters $\mu \longrightarrow \mathbb{G}_m$. In particular, X_σ is a so-called "V-manifold", i.e., isomorphic locally to a non-singular variety divided by a finite group.

§2. General embeddings

The first purpose of this section is to classify all normal equivariant embeddings of a torus T in terms of "systems" of convex rational polyhedral cones in $N_{\mathbb{R}}$, by patching what has been done in §1. Then we set up an equivariant resolution of singularities for those objects.

The first basic result we start with is the following:

Theorem 5 (Sumihiro): *Let* T *be a torus acting on a normal variety* X, *then every point of* X *admits an open invariant affine neighborhood.*

We have in fact the following lemma, in which we use the terminology of [G.I.T., Chapter 1]:

Lemma 1[1]: *Let* T *be a torus acting on a normal variety* X: *for every point* $x \in X$, *there exists an open invariant neighborhood* U *of* x *and an ample invertible sheaf* L *on* U *which admits a* T-*linearization.*

Let us show first that Lemma 1 implies Theorem 5. Let U be an invariant open neighborhood of x, and let L be a very ample invertible sheaf on U, equipped with a T-linearization: these data induce an embedding $U \xrightarrow{i} \mathbb{P}^N$, an action of T on $\mathbb{P}^N$, and a T-linearization

[1] In fact in this lemma which is also due to Sumihiro, T can be replaced by any connected linear algebraic group G; the proof we give below is valid without any modification whenever $\mathrm{Pic}(G) = 0$.

of $\mathcal{O}_{\mathbb{P}^N}(1)$ such that i is T-linear and compatible with the T-linearizations of L and $\mathcal{O}_{\mathbb{P}^N}(1)$ respectively [cf. G.I.T., p. 35] the result is then a consequence of the following lemma, applied to Z = closure of U in $\mathbb{P}^N$ and $Y = Z-U$:

Lemma 2: Let Z be a projective variety on which T acts and let L be an ample T-linearized invertible sheaf on Z. Let Y be a closed invariant subset of Z and $u \in Z-Y$: then there exists an affine invariant open set containing x and contained in Z-Y.

Proof of Lemma 2. There exists an integer N and a section s_o of L^N such that $s_o \equiv 0$ on Y and $s_o(x) \neq 0$. (Since Z is projective and L is ample, the open subsets Z_s of Z where s is non-zero, for $s \in \Gamma(L^{\alpha})$, $\alpha \geq 1$, are affine and form a basis of the topology of Z.)

Let $V \subset \Gamma(Z, L^N)$ be the subspace of sections s such that $s \equiv 0$ on Y: V is invariant under the action of T and admits a basis $\{s_\nu\}$, $(\nu = 1, \cdots, \ell)$ where each s_ν is a semi-invariant (i.e., there exists a character χ^{r_ν} such that $g.s_\nu = \chi^{r_\nu}(g).s_\nu$, all $g \in T$). The existence of s_o implies then that $s_\nu(x) \neq 0$ for some ν so, in other words, $x \in Z_{s_\nu} \subset Z-Y$, and Z_{s_ν} is invariant. q.e.d.

Proof of Lemma 1. We will use the following remark:

Lemma 3: Let X and S be two irreducible normal varieties such that S is rational. Let D be a Weil divisor on X×S: then there exist two Weil divisors D_1 and D_2 on X and S respectively such that D is linearly equivalent to $p_1^{-1}(D_1) + p_2^{-1}(D_2)$. Moreover if D is a Cartier divisor, then D_1 and D_2 are Cartier divisors.

Assuming this, let U_o be an affine neighborhood of x and let $D = X - U_o$, considered as a Weil divisor whose multiplicities all are equal to one. If $\sigma: T \times X \longrightarrow X$ is the map defining the action of T, it follows from Lemma 3 applied to σ^*D that for all $g \in T$, the transform D^g of D by g is linearly equivalent to D.

Now let U be $X - \bigcap_{g \in T} \mathrm{Supp}(D^g)$. U is an invariant open subset and it follows from the preceding remark that the restriction of D on U is a Cartier divisor; let $L = \mathcal{O}_U(D)$: by construction $|D|$ is base point free, hence L is generated by $\Gamma(L)$.

Since Pic T = 0, it follows from Lemma 3 that p_2^*L and σ^*L are isomorphic. With the notations of [G.I.T., p. 30], we want to show the existence of an isomorphism $\phi: p_2^*L \xrightarrow{\sim} \sigma^*L$ such that, on $T \times T \times U$, the following identity holds:

$$(*) \qquad p_{23}^*\phi \cdot (1_T \times \sigma)^*\phi = (\mu \times 1_U)^*\phi \quad .$$

(This is a T-linearization of L.) Let us notice first that the two members of (*) can differ only by an element of $H^o(\mathcal{O}^*_{T \times T \times U})$ and by a result of Rosenlicht (cf. loc. cit.) such an element is of the form $(g, g', x) \longmapsto \chi_1(g)\chi_2(g')h(x)$, where χ_1, χ_2 are characters on T and $h \in \Gamma(\mathcal{O}_U^*)$. Nor choose an isomorphism ϕ such that the restriction of ϕ to $\{e\} \times U$ is the canonical one: by restricting the two sides of (*) to $\{e\} \times \{e\} \times U$ we see that h = 1; then, by restrictions to $\{e\} \times G \times U$ and $G \times \{e\} \times U$, $\chi_1 = \chi_2 = 1$, and (*) holds everywhere.

It remains to show that L is ample on U. Let $x_1,\cdots,x_\ell$ be a system of coordinates for the affine variety U_o, considered as rational functions on U: there exists an integer N such that $x_1,\cdots,x_\ell$ are sections of $\mathcal{O}_U(ND) \simeq L^N$. Moreover, for all $g \in T$, $gx_1,\cdots,gx_\ell$ are sections of $\mathcal{O}_U(ND^g) \simeq L^N$ and form a system of coordinates for gU_o. The map $U \longrightarrow \mathbb{P}(\Gamma(U,L^N))$ induces therefore an immersion on each gU_o: since it is clearly injective, it is an immersion and L^N is very ample. q.e.d.

Proof of Lemma 3: Since S is rational we can find an affine open set U = Spec A in S, such that A is a localization of a polynomial ring: $k[X_1,\cdots,X_n]_f$. Let η be the generic point of X: on $\mathrm{Spec}(k(\eta))\times U$, D is the divisor of a rational function f; since the support of D-(f) avoids an open set of X×S, we must have $D-(f) = p_1^{-1}(D_1) + p_2^{-1}(D_2)$. q.e.d.

Definition 3: A finite rational partial polyhedral decomposition (we abbreviate this to f.r.p.p. decomposition) of $N_{\mathbb{R}}$ is a finite set $\{\sigma_\alpha\}$ of convex rational polyhedral cones in $N_{\mathbb{R}}$ such that:

(i) if σ is a face of σ_α, then $\sigma = \sigma_\beta$ for some β

(ii) $\forall\ \alpha,\beta$, $\sigma_\alpha \cap \sigma_\beta$ is a face of σ_α and σ_β.

Let $\{\sigma_\alpha\}$ be an f.r.p.p. decomposition of $N_{\mathbb{R}}$: in §1, Theorem 1', we associated to each σ_α an affine normal equivariant embedding

X_{σ_α} of T; patching together the X_{σ_α}'s along the $X_{\sigma_\alpha \cap \sigma_\beta}$'s (cf. Theorem 3) we get a normal scheme, which will be denoted by $X_{\{\sigma_\alpha\}}$ and which is separated (because $X_{\sigma_\alpha} \cap X_{\sigma_\beta}$ is by definition the affine scheme $X_{\sigma_\alpha \cap \sigma_\beta}$ and $\Gamma(\mathcal{O}_{X_{\sigma_\alpha \cap \sigma_\beta}})$ is generated by $\Gamma(\mathcal{O}_{X_{\sigma_\alpha}})$ and $\Gamma(\mathcal{O}_{X_{\sigma_\beta}})$, as one checks immediately).

With these notations:

Theorem 6: (i) The correspondence $\{\sigma_\alpha\} \longmapsto X_{\{\sigma_\alpha\}}$ defines a bijection between the f.r.p.p. decompositions of $N_{\mathbb{R}}$ and the isomorphy classes of equivariant normal embeddings of T.

(ii) The map $\sigma_\alpha \longmapsto X_{\sigma_\alpha}$ defines a bijection between the set of σ_α's and the set of invariant open affine subsets of $X_{\{\sigma_\alpha\}}$.

(iii) The map which associates to each X_{σ_α} its unique closed orbit, denoted by $\mathcal{O}^{\sigma_\alpha}$, defines a bijection between the set of σ_α's and the set of orbits in $X_{\{\sigma_\alpha\}}$. Moreover $\sigma_\alpha \subseteq \sigma_\beta$ if and only if $\mathcal{O}^{\sigma_\beta} \subset \overline{\mathcal{O}^{\sigma_\alpha}}$.

Proof: Let X be a normal equivariant embedding of T and let $\mathcal{O}$ be an orbit in X. Then the set

$$U_{\mathcal{O}} = \left\{ x \,\middle|\, \mathcal{O} \subset \overline{\mathcal{O}(x)} \right\}$$

is open and affine; in order to see that, let U be an open affine invariant subset of X containing Φ (which exists by Theorem 5), then $U_0 \subset U$ and we are reduced to the affine case which has been studied in Theorem 2.

Theorem 6 is then a straightforward consequence of this remark and what has been done in §1.

<u>Theorem</u> 7. <u>Let</u> $\{\sigma_\alpha\}$ <u>and</u> $\{\sigma_\beta\}$ <u>be two</u> f.r.p.p. <u>decompositions of</u> $N_{\mathbb{R}}$. <u>Then there exists a morphism</u> $X_{\{\sigma_\alpha\}} \longrightarrow X_{\{\sigma'_\beta\}}$ <u>of equivariant embeddings if and only if for all</u> α <u>there exists</u> β <u>such that</u> $\sigma_\alpha \subseteq \sigma'_\beta$.

Theorem 7 is a consequence of Theorem 3 and of the fact that if $f: X_{\{\sigma_\alpha\}} \longrightarrow X_{\{\sigma'_\beta\}}$ is a map, then $f(X_{\sigma_\alpha})$ is contained in some $X_{\sigma'_\beta}$: in fact, for all $x \in X_{\sigma_\alpha}$, $\overline{\Phi(x)} \supseteq \Phi^{\sigma_\alpha}$, hence $\overline{\Phi(f(x))} \supseteq f(\Phi^{\sigma_\alpha})$. Therefore $f(x)$ is in the set $U_{f(\Phi^{\sigma_\alpha})}$.

The proof of the next theorem is left to the reader (use the valuative criterion of properness and the description of morphisms $\operatorname{Spec} k((t)) \longrightarrow T$ used in the proof of Theorem 2,b).

Theorem 8. With the notations of Theorem 7, the map $X_{\{\sigma_\alpha\}} \longrightarrow X_{\{\sigma'_\beta\}}$ is proper if and only if $\cup\sigma_\alpha = \cup\sigma'_\beta$. Moreover $X_{\{\sigma_\alpha\}}$ is complete if and only if $\cup\sigma_\alpha = N_{\mathbb{R}}$.

Now we seek to classify the coherent sheaves of fractional ideals and the divisors on $X_{\{\sigma_\alpha\}}$ in terms of the f.r.p.p. decompositions $\{\sigma_\alpha\}$, at least those which are "invariant" under the T-action.

Let $T \overset{i}{\hookrightarrow} X$ be the canonical inclusion, then $i_*(\mathcal{O}_T)$ is a quasi-coherent $\mathcal{O}_X$-module canonically embedded in the constant sheaf $\mathbb{R}(X)$ of rational functions on X; moreover $i_*(\mathcal{O}_T)$ is equipped with a natural action of T.

Let $\mathfrak{J}$ be a coherent sheaf of fractional ideals on X, contained in $i_*(\mathcal{O}_T)$, which is T-invariant (in particular, $\mathfrak{J}|_T \cong \mathcal{O}_T$): we are going to define a map

$$\text{ord}\,\mathfrak{J}: \cup\sigma_\alpha \longrightarrow \mathbb{R}$$

in the following way:

if $a \in N \cap \cup\sigma_\alpha$, then the corresponding 1-P.S. of T extends to a map

$$\lambda_a: \text{Spec}\,k[X] \longrightarrow X_{\{\sigma_\alpha\}} \qquad \text{(cf. Theorem 1')}$$

and we put

$$\text{ord}\,\mathfrak{J}(a) = \text{ord}_0(\lambda_a^*\mathfrak{J}).$$

If we fix α such that $\lambda_a(\text{Spec}\,k[X]) \subset X_{\sigma_\alpha}$, then

$$X_{\sigma_\alpha} = \text{Spec}\,R_\alpha \quad \text{where } R_\alpha = \bigoplus_{\substack{r \in M \\ r \geq 0 \\ \text{on } \sigma_\alpha}} k.\mathfrak{X}^r \quad \text{and } \mathfrak{J}|X_{\sigma_\alpha} \text{ is given by a}$$

graded sub-k-vector space J_α of $\Gamma(\mathcal{O}_T) = \bigoplus_{r\in M} k.\mathfrak{X}^r$ which is an R_α-module: since $\mathfrak{J}$ is coherent, $J_\alpha = \sum_{i=1}^{N} R_\alpha.\mathfrak{X}^{r_i}$ and since $\Gamma(\lambda_a^*\mathfrak{J}) = \sum_i k[X].X^{\langle r_i,a\rangle}$, ord $\mathfrak{J}(a) = \min_i \langle r_i,a\rangle$. Now we extend ord $\mathfrak{J}$ to the whole of $\bigcup\sigma_\alpha$ by the formula:

$$\text{ord}\, \mathfrak{J}(x) = \min_i \langle r_i,x\rangle \ .$$

We define by this way a function $\cup\, \sigma_\alpha \longrightarrow \mathbb{R}$ which satisfies the following properties, and which depends only on $\mathfrak{J}$:

(*)

(i) ord $\mathfrak{J}(\lambda x) = \lambda.\text{ord}\,\mathfrak{J}(x)$, $\quad \lambda \in \mathbb{R}^+$

(ii) ord $\mathfrak{J}$ is continuous, piecewise-linear

(iii) ord $\mathfrak{J}$ $(N \cap \cup\sigma_\alpha) \subset \mathbb{Z}$

(iv) ord $\mathfrak{J}$ is convex on each σ_α (1)

Conversely let f be a function $\cup\sigma_\alpha \longrightarrow \mathbb{R}$ satisfying the conditions (*) and for all α put

$$(J_f)_\alpha = \bigoplus_{\substack{r\in M \\ r\geq f \text{ on } \sigma_\alpha}} k.\mathfrak{X}^\alpha$$

$(J_f)_\alpha$ is a T-invariant $\Gamma(\mathcal{O}_{X_{\sigma_\alpha}})$-module.

---

(1) We shall use later the fact that if f is a function $\cup\sigma_\alpha \longrightarrow \mathbb{R}$ satisfying the properties (*), then for each α there exist $r_i \in M$ such that $f(x) = \min_i\langle r_i,x\rangle$ for $x \in \sigma_\alpha$. This remark makes explicit the structure of the functions that we have to consider. Note that here convex means

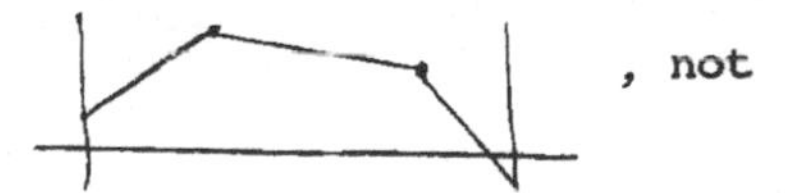 , not 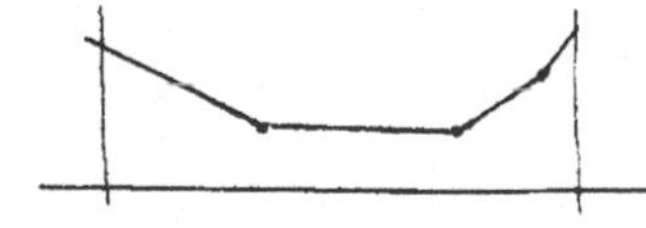

Theorem 9.

I. Let f be a function $\bigcup \sigma_\alpha \longrightarrow \mathbb{R}$ satisfying the conditions (*). Then the $\widetilde{(J_f)_\alpha}$ can be naturally patched together in a coherent sheaf $\mathfrak{J}_f$ of T-invariant fractional ideals contained in $i_*(\mathcal{O}_T)$. Moreover $\mathfrak{J}_f$ is complete[2].

II. We have the following properties:

a) $\text{ord}\ \mathfrak{J}_f = f$

b) $\mathfrak{J}_{\text{ord}\ \mathfrak{J}}$ is the completion[2] of $\mathfrak{J}$.

c) The maps $\mathfrak{J} \longmapsto \text{ord}_{\mathfrak{J}}$ and $f \longmapsto \mathfrak{J}_f$ define a bijection between the set of coherent sheaves of T-invariant complete fractional ideals and the set of functions f satisfying (*).

d) $\mathfrak{J} \subset \mathfrak{J}_f$ iff $\text{ord}\ \mathfrak{J} \geq f$

e) $\text{ord}\ \mathfrak{J}_1 \cdot \mathfrak{J}_2 = \text{ord}\ \mathfrak{J}_1 + \text{ord}\ \mathfrak{J}_2$

f) $\mathfrak{J}|X_{\sigma_\alpha} = \mathcal{O}_{X_{\sigma_\alpha}}$ iff $\text{ord}\ \mathfrak{J} \equiv 0$ on σ_α

g) $\mathfrak{J}_{f_1}$ and $\mathfrak{J}_{f_2}$ are isomorphic as $\mathcal{O}_X$-modules if and only if $f_1 - f_2$ is linear.

---

(2) If A is an integral domain, K its quotient field and J an A-module contained in K, an element z of K is said to be integrally dependent on J if it satisfies an equation of the form

$$z^q + a_1 z^{q-1} + a_2 z^{q-2} + \dots + a_q = 0, \quad a_i \in J^i.$$

J is called complete if it is equal to its integral closure in the preceding sense. For this notion, we refer the reader to Zariski-Samuel, Vol. II, Appendix 4.

III. a) $\mathfrak{J}^{-1} = \mathfrak{J}_g$ where g is the convex interpolation[3] of $-\text{ord}\ \mathfrak{J}$ on $\bigcup_\alpha Sk^1\sigma_\alpha$.

b) $(\mathfrak{J}^{-1})^{-1} = \mathfrak{J}$ if and only if $\mathfrak{J}$ is complete and ord $\mathfrak{J}$ is the convex interpolation of a function: $\bigcup_\alpha Sk^1\sigma_\alpha \longrightarrow \mathbb{Z}$. Moreover, there exists a bijective correspondence between the set of T-invariant Weil divisors and the set of integral functions on $\bigcup_\alpha Sk^1\sigma_\alpha$.

c) The following are equivalent:

i) $\mathfrak{J}$ is invertible

ii) $\mathfrak{J}\cdot\mathfrak{J}^{-1} = \mathcal{O}_X$

iii) ord $\mathfrak{J}$ is linear on each σ_α.

d) $\hat{\Omega}^n_X \cong \mathfrak{J}_k$ where k is the convex interpolation of the constant function with value -1 on $Sk^1\{\sigma_\alpha\}$.

Proof of Theorem 9.

1. Let f be a function satisfying (*).

Let σ_β be a face of σ_α, i.e., there exists an $r_o \in M$ such that $r_o \geq 0$ on σ_α and such that $\sigma_\beta = \sigma_\alpha \cap \left\{x \middle| r_o(x) = 0\right\}$. Now

---

[3] For any convex rational conical polyhedron σ in $N_{\mathbb{R}}$, $Sk^1\sigma$ is the set of primitive integral vectors in the 1-dimensional faces.

Given $h: Sk^1\sigma \longrightarrow \mathbb{Z}$, the convex interpolation of h is the least function $\tilde{h}$ on σ satisfying (*) which is $\geq$ h on $Sk^1\sigma$:

$$\tilde{h}(x) = \min_{\substack{\ell \text{ linear integral} \\ \ell \geq h \text{ on } Sk^1\sigma}} \ell(x)$$

If $\{\sigma_\alpha\}$ is an f.r.p.p. decomposition of $N_{\mathbb{R}}$, for all $h: \bigcup_\alpha Sk^1\sigma_\alpha \longrightarrow \mathbb{Z}$, the convex interpolation of h is the function $\tilde{h}$ on $\bigcup_\alpha \sigma_\alpha$ such that for each α, $\tilde{h}|\sigma_\alpha$ is the convex interpolation of $h|Sk^1\sigma_\alpha$ in the preceding sense.

$r \geq f$ on σ_β iff $r + Nr_o \geq f$ on σ_α for some N, so $(J_f)_\beta$ appears as the localization $\left((J_f)_\alpha\right)_{\mathfrak{X}^{r_o}}$ of $(J_f)_\alpha$, and the $\widetilde{(J_f)_\alpha}$'s can be patched together in a sheaf of T-invariant fractional ideals $\mathcal{J}_f$.

Now fix α and let $f(x) = \min\limits_{1 \leq i \leq p} \langle r_i, x\rangle$ on σ_α then we have clearly $\sum R_\alpha . \mathfrak{X}^{r_i} \subset (J_f)_\alpha$ and we are going to prove that in fact $(J_f)_\alpha$ is the completion of $\sum R_\alpha . \mathfrak{X}^{r_i}$:

(i) Let $r \in M$ satisfy $r \geq f$ on σ_α and assume that σ_α is defined by the inequalities $\langle \rho_j, x\rangle \geq 0$, $j = 1 \cdots q$. Then we have

$$\min_{i,j} (\langle \rho_j, x\rangle, \langle r_i - r, x\rangle) \leq 0 \qquad \text{for all } x \in N_{\mathbb{R}}$$

hence, there exist integers $n_j, m_i \geq 0$ such that

$$\sum_j n_j \rho_j + \sum_i m_i (r_i - r) = 0$$

$$\text{or} \qquad \sum_j n_j \rho_j + \sum_i m_i r_i = (\sum_i m_i) r.$$

If we put $M = \sum\limits_i m_i$, we get a relation of integral dependence for $\mathfrak{X}^r$ on $\sum R_\alpha . \mathfrak{X}^{r_i}$:

$$(\mathfrak{X}^r)^M = \mathfrak{X}^{\left(\sum\limits_j n_j \rho_j\right)} . \prod_i (\mathfrak{X}^{r_i})^{m_i} .$$

(ii) Conversely, one checks immediately that if $\mathfrak{X}^r$ is in the completion of $\sum\limits_i R_\alpha . \mathfrak{X}^{r_i}$, then $r \geq f$ and so $\mathfrak{X}^r \in (J_f)_\alpha$.

We deduce from this remark that J_f is coherent and complete.

II. a) is trivial.

b) Assume that $\Gamma(X_{\sigma_\alpha},\mathfrak{J})$ is generated over R_α by $\mathfrak{X}^{r_i}$, $i = 1,\cdots,p$, then ord $\mathfrak{J} = \min\langle r_i, x\rangle$ on σ_α and as in the proof of I, $(J_\alpha)_{\mathrm{ord}\,\mathfrak{J}}$ appears as the completion of $\sum_i R_\alpha \cdot \mathfrak{X}^{r_i} = \Gamma(X_{\sigma_\alpha},\mathfrak{J})$.

c) is a direct consequence of a) and b), and the rest of the proof of part II is straightforward.

III. a) Let $r \in M$, then $\mathfrak{X}^r \in \mathfrak{J}^{-1}\big|_{X_{\sigma_\alpha}}$ if and only if $\mathfrak{X}^r\cdot\mathfrak{J}\big|_{X_{\sigma_\alpha}} \subset \mathcal{O}_{X_{\sigma_\alpha}}$, hence if and only if $r \geq -\mathrm{ord}\,\mathfrak{J}$ on σ_α by III.d, and this last condition is satisfied if and only if $r \geq g$, where g is the convex interpolation of $-\mathrm{ord}\,\mathfrak{J}$ on $Sk^1\sigma_\alpha$.

b) The first part of the statement is a direct consequence of a). Now let D be a Weil divisor on $X = X_{\{\sigma_\alpha\}}$, and define $\mathcal{O}_X(D)$ by

$$\mathcal{O}_X(D)(U) = \left\{ f \in \mathbb{R}(X) \,\middle|\, (f) + D \geq 0 \right\}$$

for all open U in X, then one checks that a coherent sheaf of fractional ideals $\mathfrak{J}$ is of the type $\mathcal{O}_X(D)$ if and only if $(\mathfrak{J}^{-1})^{-1} = \mathfrak{J}$ (this is true on any normal variety); then the correspondence $D \longmapsto \mathrm{ord}(\mathcal{O}_X(D))$ defines an isomorphism between the group of Weil divisors $\subset X-T$ and the group of integral functions on $\bigcup_\alpha Sk^1\sigma_\alpha$.

c) follows from the preceding proof.

d) On $T = \mathrm{Spec}\, k[X_1,X_1^{-1},X_2,X_2^{-1},\cdots,X_n,X_n^{-1}]$ let $\omega = dX_1/X_1\wedge\cdots\wedge dX_n/X_n$ be the unique (up to a constant) differential form of degree n which is invariant by translations: ω has a pole of order 1 on each orbit of codimension 1.

<u>Theorem</u> 10. <u>Let</u> $X = X_{\{\sigma_\alpha\}}$ <u>and</u> $\mathfrak{J}$ <u>be a coherent sheaf of fractional ideals contained in</u> $i_*(\mathcal{O}_T)$ <u>and</u> T-<u>invariant. Let</u> $B_{\mathfrak{J}}(X)$ <u>be the</u>

<u>normalization of the variety obtained by blowing up</u> $\mathfrak{J}$.
<u>Then</u> $B_{\mathfrak{J}}(X)$, <u>as an allowable embedding of</u> T, <u>is described by the</u> f.r.p.p. <u>decomposition of</u> $N_{\mathbb{R}}$ <u>obtained by subdividing the</u> σ_α's <u>into the biggest polyhedra on which</u> ord $\mathfrak{J}$ <u>is linear</u>.

This theorem is a direct consequence of Theorem 7 and Theorem 9, using the universal property characterizing the normalization of the blowing up as the minimal normal Y with a birational map f to X such that the induced sheaf of fractional ideals $f^*\mathfrak{J}$ is invertible on Y.

<u>Theorem 11.</u> <u>Let</u> $X = X_{\{\sigma_\alpha\}}$: <u>then there exists a T-invariant sheaf of ideals</u> $\mathfrak{J} \subset \mathcal{O}_X$ <u>such that</u> $B_{\mathfrak{J}}(X)$ <u>is non singular</u>.

<u>Proof of Theorem 11.</u>

(1) Let σ be a convex rational polyhedral cone in $N_{\mathbb{R}}$: if $\sigma = \left\{ \sum \lambda_i x_i \,\middle|\, \lambda_i \geq 0 \right\}$ where the x_i are primitive vectors in the lattice N, we shall write $\sigma = \langle x_1, \cdots, x_N \rangle$. We shall say that σ is a <u>simplex</u> if $\sigma = \langle x_1, \cdots, x_N \rangle$ where the x_i are linearly independent.

Then, for a simplex $\sigma = \langle x_1, \cdots, x_N \rangle$ we introduce the <u>multiplicity</u> of σ as being the index of $\sum \mathbb{Z} x_i$ in $(\sum \mathbb{R} x_i) \cap N$, so that $\mathrm{mult}(\sigma) = 1$ if and only if σ is generated by a part of a basis of N over $\mathbb{Z}$. Moreover, if $\sigma = \langle x_1, \cdots, x_N \rangle$ where the x_i are linearly independent and primitive in N, then mult $\sigma > 1$ means that there exists $x = \alpha_1 x_1 + \cdots + \alpha_N x_N$ in N with $0 \leq \alpha_i < 1$, some $\alpha_i \neq 0$.

(2) If $f\colon \cup\, \sigma_\alpha \longrightarrow \mathbb{R}$ is a function satisfying conditions (*), then the biggest polyhedra in each σ_α on which f is linear will be called the <u>polyhedra associated to</u> f.

In order to prove Theorem 11, we need to find a function $f\colon \cup\, \sigma_\alpha \longrightarrow \mathbb{R}^+$ satisfying the conditions (*) such that the associated polyhedra are simplices of multiplicity 1.

Let us notice first of all that we can weaken the condition $f(\cup\sigma_\alpha \cap N) \subset \mathbb{Z}$: if we get a function such that $f(\cup\sigma_\alpha \cap N) \subset \mathbb{Q}$, then we can multiply f by a suitable integer in order to get the required function.

(3) We construct f by induction on the number of σ_α. So let σ_{α_o} be of maximal dimension and assume that we have found a function f_o on $\bigcup_{\alpha\neq\alpha_o} \sigma_\alpha$ satisfying the required properties: by the following lemma we get a function for $\bigcup_\alpha \sigma_\alpha$ which extends f_o, which is piecewise linear convex rational and such that the associated polyhedra are simplices:

<u>Lemma 1</u>. <u>Let</u> σ <u>be a polyhedron and</u> f_o <u>a function piecewise linear convex rational on the boundary</u> $\partial\sigma$ <u>of</u> σ <u>and let</u> $x_o \in N \cap \text{int}\ \sigma$: <u>define a function</u> f <u>on</u> σ <u>by</u>:

$$f(\alpha x_o + \beta y) = \alpha C + \beta f_o(y),\ y \in \partial\sigma\ ,\ \alpha,\beta \geq 0$$

<u>where</u> $C \in \mathbb{Q}^+$.

Then, if C is large enough, f is convex and the associated polyhedra of f are of the form $<\tau, x_o>$ where τ is associated to f_o.

(4) Now consider the set F of continuous piecewise linear rational positive functions f on $\bigcup_\alpha \sigma_\alpha$ which extend f_o and such that the associated polyhedra are simplices. For every $f \in F$ we define the *multiplicity of* f as $\underset{\tau \text{ associated to } f}{\text{Max}} (\text{Mult } \tau)$, and we have to show the existence of an f such that $\text{Mult}(f) = 1$.

But, if $f \in F$, by the following lemmas, we can decrease the number of τ associated to f such that $\text{Mult}(\tau) = \text{Mult}(f)$, and then improve the multiplicity of f.

Lemma 2. Let f be a convex piecewise linear rational positive function on a polyhedron σ *and let* $\{\tau_i\}$ *be the associated subpolyhedra to f.*

Let $x_o \in N \cap \sigma$ *and let us consider the set* T *of all* τ_i *such that* $x_o \notin \tau_i$ *and all* $<\tau_i, x_o>$ *where* $x_o \notin \tau_i$ *but* τ_i *is a face of some* τ_j *such that* $x_o \in \tau_j$.

a) T *is a f.r.p.p. decomposition covering* σ .

b) *if* $\epsilon \in \mathbb{Q}^+$ *define* $f_{x_o,\epsilon}$ *on* σ *by* $f_{x_o,\epsilon} = f$ *on* τ_i *such that* $x \notin \tau_i$; $f_{x_o,\epsilon} = f+\epsilon g$ *on* $<x_o, \tau_i>$ *where* g *is a linear function such that* $g(x_o) = 1$ *and* $g|\tau_i \equiv 0$.

Then for ϵ *sufficiently small* $f_{x_o,\epsilon}$ *is still convex and its associated polyhedra are the elements of* T.

<u>Lemma 3</u>. (i) <u>if</u> τ_1 <u>and</u> τ_2 <u>are two simplices in</u> $N_{\mathbb{R}}$ <u>such that</u> τ_1 <u>is a face of</u> τ_2, <u>then</u> Mult $\tau_1 | $ Mult τ_2.

(ii) <u>Let</u> $\tau = \langle x_1, \cdots, x_N \rangle$ <u>be a simplex such that the</u> x_i <u>are independent and primitive and let</u> $x = \alpha_1 x_1 + \cdots + \alpha_\ell x_\ell$, $0 < \alpha_i < 1$, $\ell \le N$, be primitive. <u>For</u> $1 \le i \le \ell$, <u>let</u> $\tau_i = \langle x_1, \ldots, \hat{x}_i, \ldots, x_N \rangle$, <u>then</u> $\mathrm{mult}\langle x, \tau_i \rangle = \alpha_i$ mult τ.

The proofs of Lemmas 1, 2, and 3 are left to the reader.

<u>Remark</u>: in fact we can choose the function f such that $f \equiv 0$ on the σ_α's such that the orbits O^{σ_α} are smooth (i.e., in the statement of Theorem 10, we can assume that Supp $\mathfrak{I}$ = Sing locus of $X_{\{\sigma_\alpha\}}$).

Let us illustrate what has been done so far by some examples.

Let $\sigma = \langle x_1, x_2 \rangle$ be a rational sector in $N_{\mathbb{R}} \cong \mathbb{R}^2$, so that X_σ is a general 2-dimensional normal affine embedding of $\mathbb{G}_m^2$ with an isolated singularity x. In the 2-dimensional case, we know there is a minimal resolution. To describe it from our point of view proceed as follows:

1) Let Σ = convex hull of $\sigma \cap N - (0)$

2) Let $x_1, v_1, \cdots, v_k, x_2$ be the points of N on $\partial\Sigma$ between x_1 and x_2

3) Subdivide σ by the set of rays $\mathbb{R}^+ . v_i$ and the sectors between them.

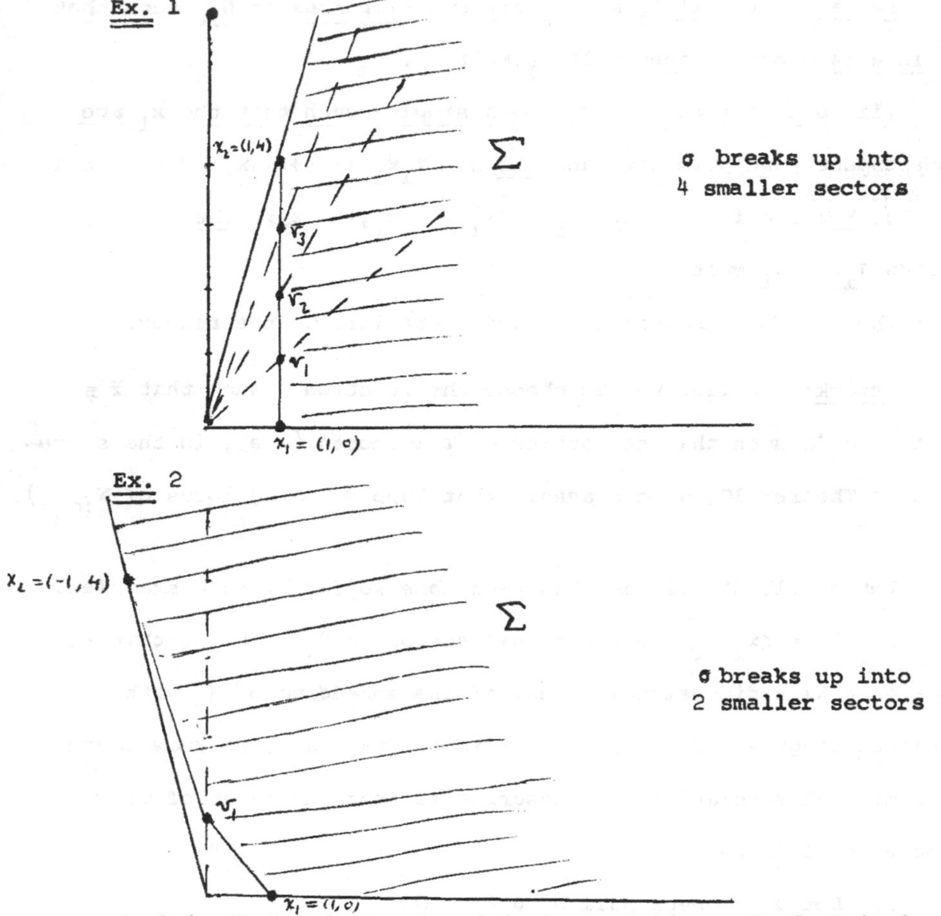

I claim that this is the minimal resolution. In fact it is a resolution because of the elementary:

<u>Lemma</u>. <u>Let</u> $\sigma = \langle x,y \rangle$, x,y <u>primitive vectors in</u> $\mathbb{Z}^2$, <u>be a sector in</u> $\mathbb{R}^2$. <u>The triangle</u> (0,x,y) <u>intersects</u> $\mathbb{Z}^2$ <u>only in its vertices if and only if</u> {x,y} <u>are a basis of</u> $\mathbb{Z}^2$.

Conversely, if $\{\tau_i\}$ defines a resolution of X_σ and if v is a vertex of Σ, then $\mathbb{R}^+ \cdot v$ = one of the τ_i. If not, let $\mathbb{R}^+ \cdot v$ fall <u>inside</u> the sector $\tau_i = \langle x_i, y_i \rangle$. Then $x_i, y_i \in \Sigma$, hence the line $\overline{x_i y_i} \subset \Sigma$, hence $v \in$ the triangle $(0, x_i, y_i)$, contradicting the lemma.

Geometrically, the resolution looks like this:

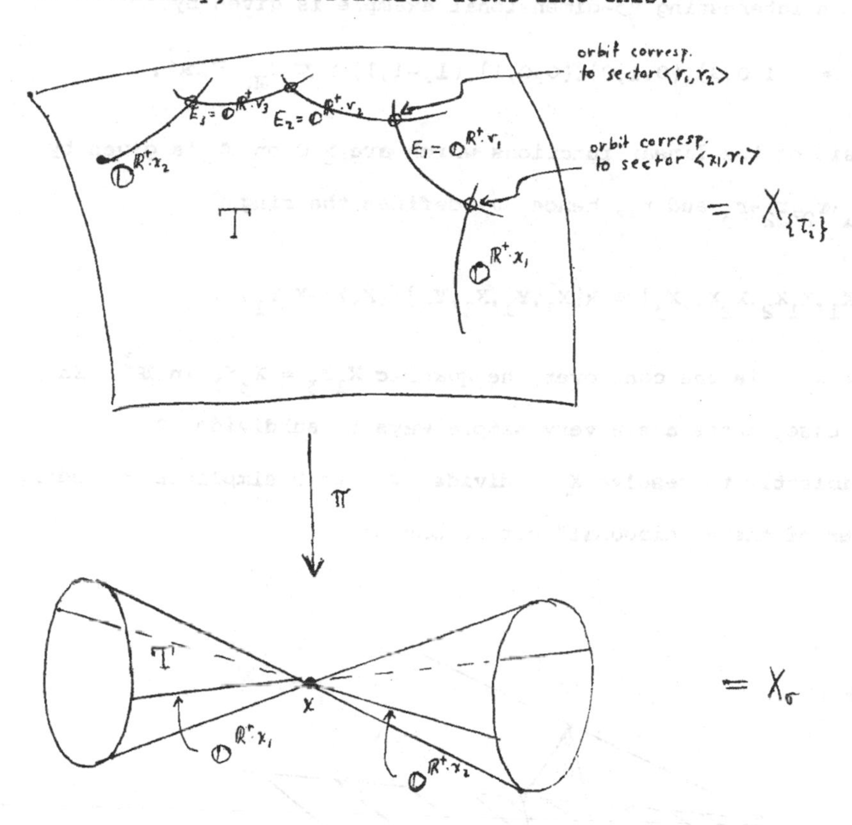

$\pi^{-1}(x)$ = chain $E_1, \cdots, E_k$ of n.s. ratℓ curves meeting transversely.

It can be shown that $(E_i^2) = -\text{mult}\langle v_{i-1}, v_{i+1}\rangle$: this is an exercise for the reader! It would be interesting to tie up this construction of the blow up more closely with its construction by continued fractions in Brieskorn, Invent. Math., 4 (1968), p. 336.

An interesting 3-dimensional example is given by

$$\sigma = \langle(1,0,0),(0,1,0),(0,0,1),(1,-1,1)\rangle \subset N_{\mathbb{R}} \cong \mathbb{R}^3.$$

A basis of the linear functions which are ≥ 0 on σ is given by r_1, r_1+r_2, r_2+r_3 and r_3, hence σ defines the ring

$$k[X_1, X_1X_2, X_2X_3, X_3] \cong k[X_1, Y_1, X_3, Y_3]/(X_1Y_3 - X_3Y_1) ,$$

hence X_σ is the cone over the quadric $X_1Y_3 = X_3Y_1$ in $\mathbb{P}^3$. In this case, there are 2 very simple ways to subdivide σ sufficiently to resolve X_σ: divide σ into 2 simplices by adding either of the 2 "diagonal" partitions δ:

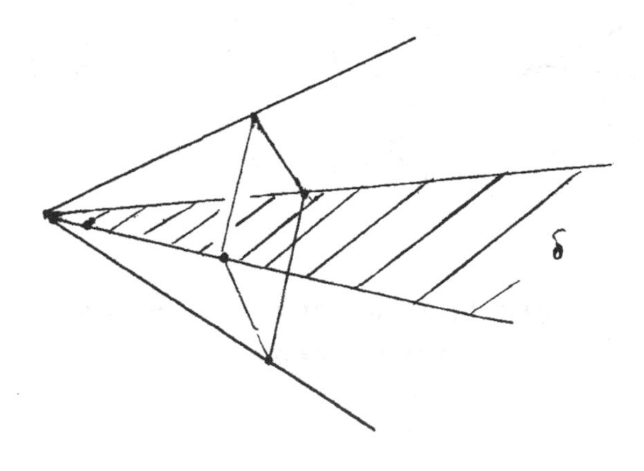

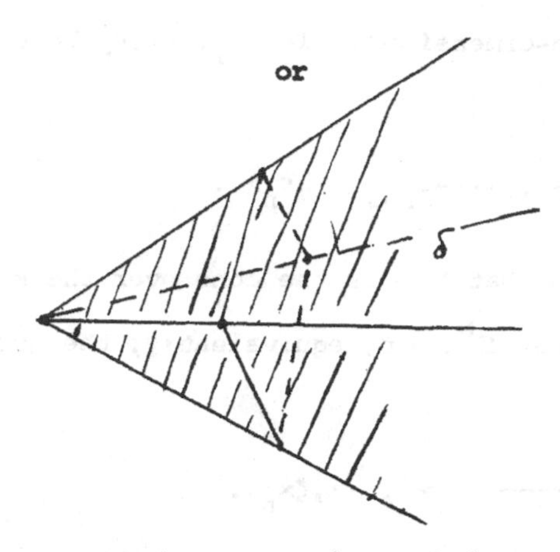

Since any 3 of the 4 vectors (1,0,0),(0,1,0),(0,0,1),(1,-1,1) form a basis of $\mathbb{Z}^3$, both of these give a n.s. model in which the singular point is replaced by $\overline{O^\delta}$, i.e., a curve isomorphic to $\mathbb{P}^1$. Moreover these 2 distinct subdivisions are associated to the convex functions:

$$f = \min(r_1, r_1+r_2); \qquad g = \min(r_1, r_3)$$

and hence the desingularizations are obtained by blowing up the ideals:

$$\mathfrak{J}_f = (X_1, Y_1) = \left(\begin{array}{l}\text{ideal of divisor which is the}\\ \text{cone over a line of the quadric}\end{array}\right)$$

$$\mathfrak{J}_g = (X_1, X_3) = \left(\begin{array}{l}\text{ideal of divisor which is the}\\ \text{cone over a line of the other ruling}\end{array}\right)$$

That these are, in fact, the 2 simplest ways to resolve a 3-dim$^{\ell}$ ordinary double point is well known.

One final example in n-dimensions: let $\epsilon_1,\cdots,\epsilon_n$ be the unit vectors in $\mathbb{Z}^n$ and let

$$\sigma = \langle \epsilon_1,\cdots,\epsilon_{n-1},-\epsilon_1-\cdots-\epsilon_{n-1} + k\epsilon_n \rangle .$$

Then it can easily be shown that X_σ is the cone over the k-tuple embedding of $\mathbb{P}^{n-1}$ in a higher $\mathbb{P}^N$, or, equivalently, the quotient of $\mathbb{A}^n$ by μ_k acting by:

$$(x_1,\cdots,x_n) \longmapsto (\zeta x_1,\cdots,\zeta x_n).$$

It is resolved by subdividing via the ray $\epsilon_n\cdot\mathbb{R}^+$ and its join to the faces of σ, this being equivalent to a simple quadratic transformation of X_σ at its singular point.

§3. Cohomology and Convexity

If X is a T-equivariant embedding of a torus T, or a T-space, let $|X|$ denote the associated rational polyhedral complex. Further, let $|f|$ represent the mapping $|X| \longrightarrow |Y|$ associated to a T-equivariant morphism $f: X \longrightarrow Y$.

Let $\mathcal{J}$ be a complete T-invariant sheaf of fractional ideals on T-space X. Set $j = \text{ord}\, \mathcal{J}$. Define the sheaf of (absolutely regular) differentials with coefficients in $\mathcal{J}$, written $\mathcal{K}_j$, by the formula :

$$\mathcal{K}_j(U) \equiv \left[\oplus k.\mathcal{X}\right], \text{where } \mathcal{X} \text{ runs through the}$$

characters with $\langle\mathcal{X},x\rangle > j(x)$ for all $x \in |U|-\{0\}$, for all T-invariant opens U in X. Clearly, $\mathcal{K}_j$ is a complete T-invariant sheaf of fractional ideals.

If $\mathcal{J} = \mathcal{O}_X$, then $j=0$ and $\mathcal{K}_0 = \hat{\Omega}_X$ because $\hat{\Omega}_X = \mathcal{F}_\delta$ where δ is the convex interpolation of 1 on the one-skeleton of $|X|$. Also, let $\mu: \overline{X} \longrightarrow X$ be a T-invariant morphism. Let $\overline{j} = j\cdot|\mu|$. Then, μ^* (section of $\mathcal{K}_j$) is a section of $\mathcal{K}_{\overline{j}}$. Furthermore, if μ is proper, we have $\mu_*\mathcal{K}_{\overline{j}}$ is exactly $\mathcal{K}_j$.

Let $\mathcal{J}$ be a T-invariant sheaf of fractional ideals on a T-space X. As $\mathcal{J} = \oplus\, \mathcal{J}^{\mathcal{X}}$, we have that $H^i(X,\mathcal{J}) = \oplus\, H^i(X,\mathcal{J}^{\mathcal{X}})$. Obviously, $H^i(X,\mathcal{J}^{\mathcal{X}}) = H^i(X,\mathcal{J})^{\mathcal{X}}$. Therefore, the cohomology $H^i(X,\mathcal{J})$ can be computed by

Theorem. For any character χ of T, we have

$$H^i(X,\mathcal{J})^{\chi} \approx H^i_A(|X|,k)$$

where A is one of the following closed subsets of $|X|$.

a) $A = |U|$ where U is the largest T-invariant open in X such that $\chi \in \mathcal{J}(U)$.

b) In case $\mathcal{J}$ is complete,

$$A = \{x \in |X| \mid \langle \chi,x\rangle \geq \text{ord } \mathcal{J}(x)\}.$$

Furthermore, if $j = \text{ord}(\mathcal{J})$, the cohomology of the differentials $\mathcal{K}_j$ is given by the formula

$$H^i(X, \mathcal{K}_j)^{\chi} \approx H^i(|X|, B^* ,k)$$

where $B = \{x \in |X| \mid \langle \chi,x\rangle \leq j(x)\}$, $B^* = B - B\cap(\text{open unit ball in } N_{\mathbb{R}})$.

Proof. I) Assume X is affine. Then

$$H^i(X,\mathcal{J}) = 0 \quad \text{for} \quad i > 0 \quad \text{and}$$

$$H^0(X,\mathcal{J})^{\chi} = \begin{cases} k.\chi & \text{if } \chi \in \mathcal{J}(X) \\ 0 & \text{otherwise.} \end{cases}$$

On the other hand, both of the opens $|X|-A$ are convex. $|X|-A$ is empty if and only if $\chi \in \mathcal{J}(X)$. The long exact sequence

$$0 \longrightarrow H^0_A(|X|,k) \longrightarrow H^0(|X|,k) \longrightarrow H^0(|X|-A,k) \longrightarrow H^1_A(|X|,k) \longrightarrow \cdots$$

has zero at the dots because $|X|$ and $|X|-A$ are contractible. Therefore, $H^i_A(|X|,k) = 0$ for $i > 0$ and $H^0_A(|X|,k) = \begin{cases} k & \text{if } A = |X| \\ 0 & \text{otherwise} \end{cases}$.

This completes the affine case of the first isomorphism.

II). Let $\{U_i\}$ be a covering of X by T-invariant open affines. Let σ_i be the simplex $|U_i|$. Intersections of U_i's are open affines and these correspond to intersections of the σ_i's. By Part I, $H^i_{\cap\sigma_j\cap A}(\cap\sigma_j,k) = 0$ if $i > 0$. Hence by Leray's theorem,

$$H^i_A(|X|,k) \approx \hat{H}^i_A(\{\sigma_j\},k).$$

Also Serre's theorem says that

$$H^i(X,\mathcal{J})^* \approx \hat{H}^i(\{U_j\},\mathcal{J})^* .$$

Again by part I, the two Čech complexes

$$\hat{C}^i_A(\{\sigma_j\},k) \quad \text{and} \quad \hat{C}^i(\{U_j\},\mathcal{J})^*$$

are isomorphic. Thus

$$\hat{H}^j_A(\{\sigma_i\},k) \approx \hat{H}^j(\{U_i\},\mathcal{J})^* .$$

This completes the proof of the first isomorphism. The proof of the second isomorphism is almost word for word the same. Q.E.D.

Essentially this proof was given by Demazure in the case he was studying; i.e., X smooth and $\mathcal{J}$ invertible.

Let $\varphi: X \longrightarrow Y$ be a proper T-equivariant morphism. Let $\mathcal{G} = \mathcal{F}_f$ be a complete sheaf on Y for some $f: |Y| \longrightarrow \mathbb{R}^+$. Denote $f \cdot |\varphi|$ by g. Then we have a φ-homomorphism $\mathcal{F}_f \longrightarrow \mathcal{F}_g$.

Corollary 1. In the above situation,

a) $H^i(X, \mathcal{F}_g) \longrightarrow H^i(Y, \mathcal{F}_f)$ is an isomorphism for all i.

b) $H^i(X, \mathcal{K}_g) \longrightarrow H^i(Y, \mathcal{K}_f)$ is an isomorphism for all i.

c) $R^i\varphi_* \mathcal{F}_g = 0$ for all $i > 0$

and $\varphi_* \mathcal{F}_g \approx \mathcal{F}_f$

and d) $R^i\varphi_* \mathcal{K}_g = 0$ for all $i > 0$

and $\varphi_* \mathcal{K}_g \approx \mathcal{K}_f$.

Proof. $|\varphi|: |X| \longrightarrow |Y|$ is a homeomorphism. The second A and the B in the theorem for X and Y are equivalent under $|\varphi|$. The theorem says that the cohomology groups are topological invariants of X, A and B. This proves a) and b). Statements c) and d) follow formally from a) and b) when Y is affine. Q.E.D.

Actually, b) and d) are not unusual in characteristic zero.*

Corollary 2. Let X be a T-space and $\mathcal{G}$ be a complete T-invariant sheaf of fractional ideals. Set $f = \text{ord}\, \mathcal{G}$. Assume $|X|$ is convex and f is convex. Then, $H^i(X, \mathcal{G})$ is zero for all $i > 0$.

---

* H. Grauert and O. Riemanschneider, "Verschwindungssätze für analytische Kohomologiegruppen auf komplexen Räumen", Inventiones mathematicae, Vol. 11, 1970, pp. 263-292.

Proof. For a given character χ, let A be given by part b) of Theorem 12. By assumption, $|X|$ and $|X|-A$ are contractible if $|X| \neq A$. The conclusion follows from the long exact sequence in the proof of the theorem. Q.E.D.

First, let's do a numerical corollary of the theorem.

Corollary 3. Let $\mathcal{J} = \mathcal{J}_f$ be a complete T-invariant sheaf of ideals on an affine T-space X. Assume that the support of $\mathcal{O}_X/\mathcal{J}$ is the zero-dimensional orbit. Then, the function $\dim \Gamma(X, \mathcal{O}_X/\mathcal{J}_{nf})$ is a polynomial for all integers $n \geq 0$. The degree of this polynomial is the dimension of X.

Proof. Let k be the dimension of X. Let Y be the normalization of X after $\mathcal{J}$ is blown up. The morphism $\mu: Y \longrightarrow X$ is proper. We have $\mu^*\mathcal{J}_{nf} \approx (\mu^*\mathcal{J}_f)^{\otimes n}$ as $\mu^*\mathcal{J}_f \equiv \mathcal{L}$ is an invertible ideal. Let F be the closed subscheme of Y defined by $\mathcal{L}$. Set $K = \mathcal{L}/\mathcal{L}^2$, which is an ample invertible sheaf on F. As F is projective of dimension k-1,

$$\chi(F, K^{\otimes n}) = \sum (-1)^i h^i(F, K^{\otimes n})$$

is a polynomial for all n and has degree k-1 by Serre's form of Hilbert's theorem.

Claim. $\chi(F, K^{\otimes n}) = \dim \Gamma(X, \mathcal{J}_{nf}/\mathcal{J}_{(n+1)f})$ for $n \geq 0$. This claim implies the corollary. For,

$$\dim \Gamma(X, \mathcal{O}_X/\mathcal{J}_{nf}) = \sum_{n > i \geq 0} \dim \Gamma(X, \mathcal{J}_{if}/\mathcal{J}_{(i+1)f})$$

$$= \sum_{n > i \geq 0} \chi(F, K^{\otimes i}).$$

To prove the claim, consider the direct images by μ of the exact sequence

$$0 \longrightarrow \mathcal{L}^{\otimes n+1} \longrightarrow \mathcal{L}^{\otimes n} \longrightarrow K^{\otimes n} \longrightarrow 0.$$

By the first corollary, $\mathcal{L}^{\otimes n}$'s have no higher direct images. Hence, $K^{\otimes n}$ has none. Furthermore, the sequence

$$\begin{array}{ccccccc} 0 \longrightarrow & \mu_*(\mathcal{L}^{\otimes n+1}) & \longrightarrow & \mu_*(\mathcal{L}^{\otimes n}) & \longrightarrow & \mu_*(K^{\otimes n}) & \longrightarrow 0 \\ & \wr\wr & & \wr\wr & & & \\ & \mathcal{J}_{(n+1)f} & & \mathcal{J}_{nf} & & & \end{array}$$

is exact. Thus, by Leray's spectral sequence,

$$H^i(F, K^{\otimes n}) \cong H^i(X, \mathcal{J}_{nf}/\mathcal{J}_{(n+1)f}).$$

These last groups are zero if $i > 0$. Therefore,

$$\chi(F, K^{\otimes n}) = h^0(F, K^{\otimes n}) = \dim \Gamma(X, \mathcal{J}_{nf}/\mathcal{J}_{(n+1)f}). \qquad \text{Q.E.D.}$$

This corollary can be proved by elementary methods. One may check that the leading coefficient of the polynomial is the volume of the set,

$$\left\{ \mathfrak{X} \in M_{\mathbb{R}} \;\middle|\; \begin{array}{l} \langle \mathfrak{X}, \lambda\rangle \geq 0 \quad \text{for all} \quad \lambda \in |X| \\ \text{but} \quad \langle \mathfrak{X}, \lambda\rangle < f(\lambda) \quad \text{for some} \quad \lambda \in |X| \end{array} \right\}$$

$$= |X|^{\vee} - \begin{bmatrix} \text{convex hull of points } \alpha \in M \\ \text{such that } \mathfrak{X}^{\alpha} \in \Gamma(|X|, \mathcal{J}) \end{bmatrix}$$

Now

$$\mathcal{F}_{nf} = \text{completion of } \mathcal{J}^n,$$

and it is well-known * the the leading coefficients of the polynomials

$\dim \mathcal{O}/\mathcal{F}_{nf}$, $\dim \mathcal{O}/\mathcal{J}^n$ are equal. Thus k! times this volume is $e(\mathcal{J})$,

the multiplicity of $\mathcal{J}$ at the fixed point of T.

Before studying the singularities of T-spaces further, let's consider some of the geometric meaning of the convexity condition in Corollary 2.

Lemma. Let X be a T-space.

a) $|X|$ is convex <===> X is proper over an affine.

b) A complete T-sheaf of fractional ideals on X, $\mathcal{J} = \mathcal{F}_f$,

is generated by its global sections if and only if

$$f = \underset{\chi \geq f \text{ on } |X|}{\text{Inf.}} \chi .$$

In this case, f is convex.

Proof. a) Let $\lambda : \mathbb{G}_m \longrightarrow T$ be a 1-P. S. of T. Assume that X is proper over an affine. Then, $\lim_{t \to 0} \lambda(t)$ exists in X if and only if, for all $f \in \Gamma(X, \mathcal{O}_X)$, $f \circ \lambda$ is regular at zero.

Therefore $\lambda \in |X| \cap N$ if and only if $\langle \alpha, \lambda \rangle \geq 0$ for all $\alpha \in |X|^\vee \cap M$. Therefore $|X| = |X|^{\vee\vee}$, hence $|X|$ is convex. To prove the converse, let

$$Y = \operatorname{Spec} k[\ldots, \chi^\alpha, \ldots]_{\alpha \in M \cap |X|^\vee}$$

and note that there is a canonical map $f: X \longrightarrow Y$. It follows easily that f is proper using the valuative criterion and the description of morphisms $g : \operatorname{Spec} k((t)) \longrightarrow T$ given in the proof of Theorem 2

*) see page 48.

b) Obvious Q. E. D.

Theorem 13. Let X be a T-space and $\mathcal{L} = \mathfrak{F}_f$ be an invertible T-invariant sheaf. Then, $\mathcal{L}$ is ample if and only if f is strictly convex in the sense that if $\{\sigma_\alpha\}$ are the polyhedra in $|X|$, then, for each σ_α, there exists a character $\mathfrak{X}$ and a positive integer n, for which a) $\mathfrak{X} \geq nf$ on $|X|$

and b) $\sigma_\alpha = \{x \in |X| \mid \langle \mathfrak{X}, x \rangle = nf(x)\}$.

Proof. $\mathcal{L}$ is ample if and only if there is an $n > 0$ and characters $\{\mathfrak{X}_\beta\}$ such that a) the $\mathfrak{X}_\beta$'s are sections of $\mathcal{L}^{\otimes(n_\beta)}$ and b) the open U_β, where $\mathfrak{X}_\beta$ generates $\mathcal{L}^{\otimes(n_\beta)}$, form an affine cover of x. But then any invariant affine U must be contained in one of the U_β's hence there will also be a section $\mathfrak{X}' . \mathfrak{X}_\beta^m$ of $\mathcal{L}^{\otimes n'}$ such that U is the open set where this generates $\mathcal{L}^{\otimes n'}$. Since the sub sets $|U|$ for such U, and σ_α in $|X|$ are the same, the theorem is a direct translation of this criterion. Q. E. D.

---

* Let R be an integrally closed domain (which is universally Japanese), $I \subset R$ an ideal and consider the subring

$$S = \bigoplus_{i=0}^{\infty} I^n . t^n \subset R[t]$$

The integral closure S' of S in $R[t]$ is equal to:

$$S' = \bigoplus_{i=0}^{\infty} (I^n)' . t^n, \quad (I^n)' = \text{completion of } I^n.$$

Then S' is a finitely generated S - module, which implies

$$I^{n-n_0} . (I^{n_0})' = (I^n)'$$

for some n_0 and all $n \geq n_0$. Thus $(I^n)'$ is a "Stable I-filtration" and the result follows: cf. Atiyah-MacDonald, Commutative Algebra, p. 118.

The reader interested in very ample sheaves should consult Demazure's paper. He has proved that ample implies very ample on a smooth complete T-space.

Generalities on Rational Resolutions

Let $f: X \longrightarrow S$ be a proper morphism between smooth varieties. Denote the dimensions of X and S by x and s. Let $\mathcal{L}$ be an invertible sheaf on X. A special case of the duality theorem for the proper morphism f is

Theorem. Assume $R^i f_*(\Omega_X^x \otimes \mathcal{L}^{\otimes -1}) = 0$ for $i > 0$. Then there are natural isomorphisms

$$R^i f_* \mathcal{L} \approx \underset{\sim}{\mathrm{Ext}}_{\mathcal{O}_S}^{s-x+i}(f_*(\Omega_X^x \otimes \mathcal{L}^{\otimes -1}), \Omega_S^s).$$

Recall that a coherent sheaf F on S is called Cohen-Macaulay of pure dimension k if and only if $\underset{\sim}{\mathrm{Ext}}_{\mathcal{O}_S}^{s-j}(F, \Omega_S^s)$ is zero unless $j = k \equiv$ dimension of the support of F. This fact together with the theorem immediately implies

Corollary. Assume $R^i f_*(\Omega_X^x \otimes \mathcal{L}^{\otimes -1})$ and $R^i f_* \mathcal{L}$ are zero for $i > 0$. Then, $f_*(\Omega_X^x \otimes \mathcal{L}^{\otimes -1})$ and $f_* \mathcal{L}$ are Cohen-Macaulay of pure dimension x. In fact, $\underset{\sim}{\mathrm{Ext}}_{\mathcal{O}_S}^{s-x}(-, \Omega_S^s)$ interchanges them.

Let's call one of these sheaves the Ext-dual of the other.

Let $g: X \longrightarrow Y$ be a proper birational morphism with X smooth. Call g a resolution (of the singularities) of Y. Define such a morphism g to be a rational resolution if

a) Y is normal; i.e., $\mathcal{O}_Y \longrightarrow g_*\mathcal{O}_X$ is an isomorphism,

b) $R^i g_* \mathcal{O}_X$ is zero for all $i > 0$,

and c) $R^i g_* \Omega^x_X$ is zero for all $i > 0$.

Remark. Condition c) is always satisfied in characteristic zero by a generalization of Kodaira's vanishing theorem [see Grauert-Riemenschneider].

Consider the two more niceness conditions on the singularities of Y:

d) $\mathcal{O}_Y$ is Cohen-Macaulay,

e) the natural homomorphism $g_*\Omega^x_X \longrightarrow \omega_Y$ is an isomorphism where ω_Y is the sheaf on Y, which is isomorphic to $\mathcal{E}xt^{s-x}_{\mathcal{O}_S}(\mathcal{O}_Y, \mathcal{O}_S)$ if Y were embedded in a smooth S.

Also, if Y is a normal variety, ω_Y is isomorphic to the double dual of the highest differentials, Ω^y_Y [see G-R].

Proposition.* Assuming the above condition c), we have that a) and b) are equivalent to d) and e).

---

*When dim Y = 2, this was proven by H. Laufer, "On rational singularities", American Journal of Mathematics, Vol. 94, 1972, pp. 597-608.

<u>Proof</u>. The problem is local on Y. So, assume Y is embedded of a smooth variety S by i. Set $f = i \cdot g$. As condition c) is verified, the theorem gives us isomorphisms

$$\oplus \quad i_* R^i g_* \mathcal{O}_X \approx \underset{\sim}{\mathrm{Ext}}^{s-x+i}_{\mathcal{O}_S}(f_* \Omega^x_X, \Omega^s_S).$$

Thus, b) is true if and only if $f_*\Omega_X$ or $g_*\Omega_X$ is Cohen-Macaulay. If a) and b), we have that $\mathcal{O}_Y \approx g_*\mathcal{O}_X$ is the Ext-dual of the Cohen-Macaulay sheaf $f_*\Omega^x_X$. Thus, $\mathcal{O}_Y$ is Cohen-Macaulay and the homomorphism in e) is the Ext-dual of the isomorphism $\mathcal{O}_Y \longrightarrow g_*\mathcal{O}_X$. Hence, a) + b) $\Longrightarrow$ d) + e).

Assume d) and e) are verified. Then, $\mathcal{O}_Y$ is Cohen-Macaulay with dual Cohen-Macaulay sheaf ω_Y by d). By e), we have that $g_*\Omega^x_X$ is Cohen-Macaulay. Equation $\oplus$ implies condition b) and that $g_*\mathcal{O}_X$ is the Ext-dual of $g_*\Omega^x_X$. Further, the Ext-dual of the homomorphism $\mathcal{O}_Y \longrightarrow g_*\mathcal{O}_X$ of Cohen-Macaulay sheaves is an isomorphism by e. Thus, $\mathcal{O}_Y \longrightarrow g_*\mathcal{O}_X$ is an isomorphism. Hence, d) + e) $\Longrightarrow$ a) + b). Q.E.D.

The forward implication gives an easy way to check Cohen-Macaulayness and to find Ext-duals. In characteristic zero, when Y is normal, Grauert and Riemenschneider have given an intrinsic characterization of the sheaf $g_*\Omega^x_X$ which is independent of the resolution g. In this case, the conditions d) and e) essentially deal only with Y. One says that Y has rational singularities if these conditions d) and e) are verified.

Returning to T-space, we can apply the above theory to prove

Theorem 14. a)* Any T-space Y is Cohen-Macualay.

b) ω_Y is $\hat{\Omega}_Y$.

c) Any proper T-equivariant resolution of the singularities of Y is rational.

Proof. By Theorem 11, there is a proper T-equivariant morphism $f: X \longrightarrow Y$ with X smooth. By Corollary 1 of Theorem 12, the condition b) and c) in the definition of rational resolutions are verified. Thus, the third statement in the theorem follows because we are dealing only with normal T-spaces. The last general theorem shows that the first two statements follow from the third. Q.E.D.

* This fact is originally proved by M. Hochster, "Rings of invariants of tori, Cohen-Macaulay rings generated by monomials, and polytopes", Annals of Mathematics, Vol. 96, 1972, pp. 318-337.

Chapter II

Semi-stable reduction

We have 2 goals in this chapter: the first is to generalize the set-up of Chapter I to embeddings of non-singular varieties U in normal varieties X such that formally at each $x \in X-U$ the pair (X,U) is isomorphic to $(\overline{T},T)$ for some torus T and equivariant $T \subset \overline{T}$. This allows us to apply an analysis similar to that of Chapter I to a much greater range of examples. In particular we want then to apply the theory to prove:

Semi-stable reduction theorem (Algebraic case): Assume char $k = 0$. Let C be a non-singular curve, $0 \in C$ a point and

$$f: X \longrightarrow C$$

a morphism of a variety X onto C such that

$$\text{res } f: X-f^{-1}(0) \longrightarrow C-\{0\}$$

is smooth. Then there is a finite morphism

$$\pi: C' \longrightarrow C$$

C' non-singular, $\pi^{-1}(0) = \{0'\}$, and a proper morphism p as follows

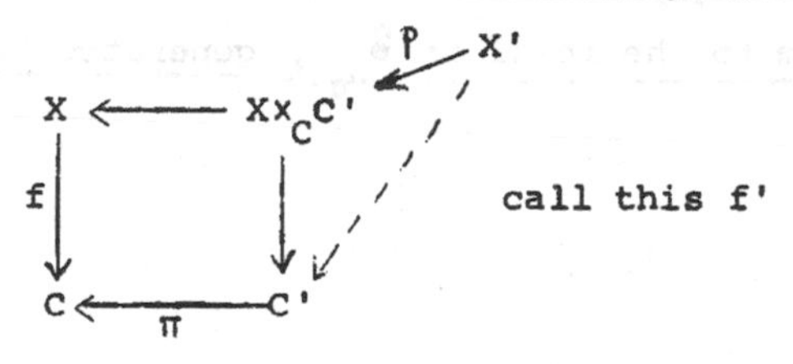

such that

a) p is an isomorphism over $C'-\{0'\}$

b) p is projective; in fact p is obtained by blowing up a sheaf of ideals $\mathcal{J}$ with $\mathcal{J}\big|_{C'-\{0'\}} \cong \mathcal{O}_{X\times_C C'}\big|_{C'-\{0'\}}$.

c) X' is non-singular, and the fibre $f'^{-1}(0')$ is reduced, with non-singular components crossing normally.

As we will see, everything here is an easy consequence of Hironaka's resolution theorems except for "reduced"!

§1. Toroidal embeddings - definitions

Definition 1. Let X be a normal variety of dimension n, U a smooth Zariski open set of X. We say that $U \subset X$ is a toroidal embedding if for every closed point x in X there exists an n-dimensional torus T, an equivariant affine embedding X_σ of T [cf. Chap. I, §1], a point t in X_σ and an isomorphism of k-local algebras:

$$\hat{\mathcal{O}}_{X,x} \xrightarrow{\approx} \hat{\mathcal{O}}_{X_\sigma,t}$$

such that the ideal in $\hat{\mathcal{O}}_{X,x}$ generated by the ideal of $X-U$ corresponds under this isomorphism to the ideal in $\hat{\mathcal{O}}_{X_\sigma,t}$ generated by the ideal of $X_\sigma-T$.

By restricting X_σ if necessary, we can assume that the orbit of t is closed in X_σ: such an equivariant affine embedding plus a formal isomorphism as above will be called a <u>local model</u> at x.

Notice that it follows formally from this definition and from the fact that X_σ-T is purely 1-codimensional, that X-U is also purely 1-codimensional [cf. EGA.IV.7.1.3]: we shall write $X-U = \bigcup_{i\in I} E_i$ where the E_i's are irreducible subvarieties of dimension n-1.

To get an idea of this definition, let us assume for a moment that dim X = 2: in that case the E_i's are curves and let x be a point in $\bigcup_{i\in I} E_i$. Then:

either a) x is a non-singular point of $\bigcup E_i$: i.e., x is in only one E_i and is non-singular on it. Then in our local model t is in a dim. 1 orbit, X_σ is non-singular at t, hence X is non-singular at x,

or b) x is singular on $\bigcup E_i$: then {t} is a dim. 0 orbit, $(X_\sigma - T)_{red}$ has an ordinary double point at t, hence x either (b_1) belongs to two components E_{i_1} and E_{i_2} which are non-singular at x and meet transversely, or (b_2) belongs to one E_i with an ordinary double point at x. Moreover the singularity of X at x is quite elementary: i.e., formally $X \cong \mathbb{A}^2/$(cyclic gp.) which the 2 branches of $\bigcup E_i$ being the image of the 2 coordinate axes.

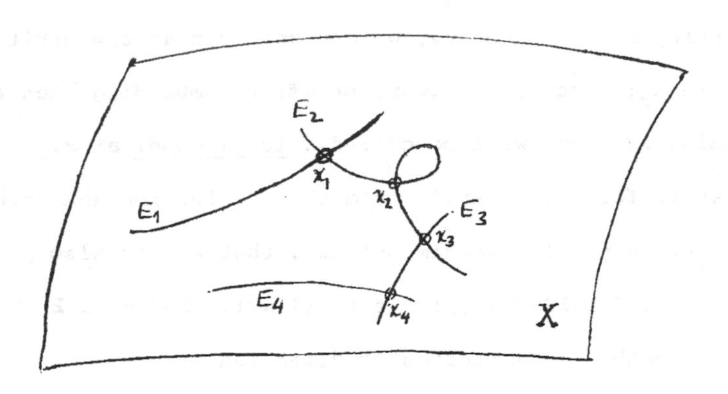

We can associate to this situation a graph, by attaching a vertex to each E_i, an edge between two vertices corresponding to a singularity of type (b_1) and a loop through a vertex corresponding to a singularity of type (b_2):

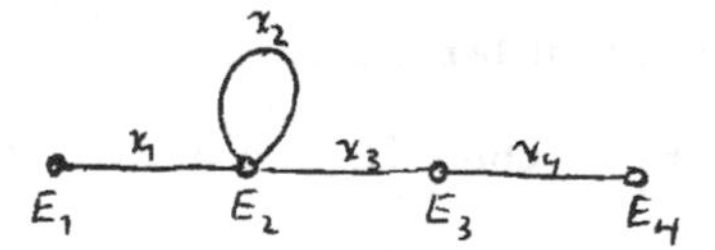

We define also a stratification[(1)] on X as follows:

i) the set U

ii) the connected components of the sets E_i - double pts. of $\bigcup E_j$

iii) the double points of $\bigcup E_i$.

[(1)] A stratification on a variety X is a finite set $S_1, \cdots, S_N$ of locally closed subsets, called the strata, such that every point of X is in exactly one stratum and such that the closure of a stratum is a union of strata.

If Y is a stratum, we define $\text{Star } Y = \bigcup_{\substack{\text{strata } Z \\ \text{such that } Y \subset \bar{Z}}} Z = X - \bigcup_{\substack{\text{strata } Z \\ \text{such that} \\ \bar{Z} \cap Y = \emptyset}} Z$

Note that Star Y is an open set.

Now, we go back to our general situation described in Def. 1: we shall restrict ourselves to the case where the E_i are normal varieties and we shall say in this case that $U \subset X$ is a toroidal embedding without self-intersection.

Proposition-Definition 2: Let $U \subset X$ be a toroidal embedding without self-intersection. Then for any subset J of I,

a) $\bigcap_{i \in J} E_i$ is normal

b) $\bigcap_{i \in J} E_i - \bigcup_{i \notin J} E_i$ is non-singular.

The components of the sets $\bigcap_{i \in J} E_i - \bigcup_{i \notin J} E_i$ define a stratification of X; moreover the components of $\bigcap_{i \in J} E_i$ are the closures of the strata. Finally if $x \in X$ and (X_σ, t) is a local model at x, then the closures $\overline{Z}$ of the strata Z, with $x \in \overline{Z}$, correspond formally to the closures of the orbits in X_σ; in particular if $x \in$ Stratum Y, then Y corresponds formally to $\mathcal{O}(t)$ itself.

Proof: Let $\mathcal{O}_\nu$, $\nu \in I'$, be the orbits of codimension 1 in X_σ, corresponding to the vertices of σ [I, Th. 2]. Then the sets $\overline{\mathcal{O}}_\nu$ are normal [I, Prop. 2] and if we denote by $\hat{\overline{\mathcal{O}}}_\nu$ the inverse image of $\overline{\mathcal{O}}_\nu$ in Spec $\hat{\mathcal{O}}_{X_\sigma, t}$, then $\hat{\overline{\mathcal{O}}}_\nu$ is still reduced, irreducible and normal (Z-S, vol. II, Ch. 8), hence are the analytic branches of $X_\sigma - T$.

On the other hand, let $I_x = \{ i \in I \mid x \in E_i \}$. Then since the E_i are assumed normal, the $\hat{E}_i$, $i \in I_x$, are the analytic branches of $X-U$ at x.

Hence the formal isomorphism Spec $\hat{\mathcal{O}}_{X,x} \xrightarrow{\approx}$ Spec $\hat{\mathcal{O}}_{X_\sigma,t}$ induces an isomorphism of I_x and I' under which the $\hat{E}_i$ correspond to the $\hat{\mathcal{O}}_\nu$. Now if J is any subset of I_x and J corresponds to $J' \subseteq I'$, let τ be the smallest face of σ containing the vertices $\nu, \nu' \in J'$. Then it follows from [I, Thm. 2] that

$$\overline{\mathcal{O}^\tau} = \bigcap_{\nu \in J'} \overline{\mathcal{O}_\nu}.$$

Since we know $\overline{\mathcal{O}^\tau}$ is normal, it follows that $\hat{\overline{\mathcal{O}^\tau}}$ is normal and corresponds formally to $\bigcap_{i\in J} \hat{E}_i$. Therefore $\bigcap_{i\in J} E_i$ is normal at x. Since x is arbitrary, $\bigcap_{i\in J} E_i$ is normal. Moreover, if we take $J = I_x$, $J' = I'$, then

$$\mathcal{O}(t) = \bigcap_{\nu\in I'} \overline{\mathcal{O}_\nu}$$

and $\mathcal{O}(t)$ is non-singular. It follows then that $\bigcap_{i\in I_x} E_i$ is non-singular at x. Since x is an arbitrary point of $\bigcap_{i\in I_x} E_i - \bigcup_{i\notin I_x} E_i$, it follows that this set is non-singular.

Call the components of the sets $\bigcap_{i\in J} E_i - \bigcup_{i\notin J} E_i$ <u>strata</u>. Obviously X is the disjoint union of these sets and their closures are just the components of the sets $\bigcap_{i\in J} E_i$ and we have checked that these do correspond formally to the closures of orbits. It remains to check the axiom of the frontier: if Y,Z are strata, $x \in Y \cap \overline{Z}$, then $Y \subset \overline{Z}$. But it suffices to prove that

(*) $$I(Y)\cdot\hat{\mathcal{O}}_{X,x} \supseteq I(\overline{Z})\cdot\hat{\mathcal{O}}_{X,x}.$$

In our local model, Y corresponds to $\mathcal{O}(t)$ and $\overline{Z}$ to $\overline{\mathcal{O}^{\tau}}$, some face τ of σ. Since $\mathcal{O}(t) \subseteq \overline{\mathcal{O}^{\tau}}$, it follows that

$$I(\mathcal{O}(t))\cdot\hat{\mathcal{O}}_{X_\sigma,t} \supseteq I(\overline{\mathcal{O}^{\tau}})\cdot\hat{\mathcal{O}}_{X_\sigma,t}$$

which proves (*). QED

Note: If we do not assume that $U \subset X$ is without self-intersection, it is still possible to define a stratification such that for $x \in X$, the stratum containing x is formally isomorphic to the orbit of t as follows:

If x is r_i-fold on E_i, $i \in I_x$, $1 \leq r_i < \infty$ then the stratum containing x is the connected component through x of the subset of $\bigcap_{i\in I_x} E_i - \bigcup_{i\notin I_x} E_i$ where the multiplicity along each E_i is equal to r_i.

Definition 3: Let Y be a stratum.

M^Y = group of Cartier divisors on Star(Y), supported on Star(Y) − U.

$M^Y_{\mathbb{R}} = M^Y \otimes \mathbb{R}$

$N^Y = \mathrm{Hom}(M^Y,\mathbb{Z})$

$N^Y_{\mathbb{R}} = N^Y \otimes \mathbb{R} = \mathrm{Hom}(M^Y_{\mathbb{R}},\mathbb{R})$

M^Y_+ = sub-semi-group of M^Y of effective divisors

$\sigma^Y = \left\{x \in N^Y_{\mathbb{R}} \,\middle|\, \langle D,x\rangle \geq 0, \text{ all } D \in M^Y_+\right\} \subset N^Y_{\mathbb{R}}.$

If Star $Y - U = \bigcup_{i \in J} E_i$, then M^Y is a subgroup of the free abelian group of Weil divisors $\sum_{i \in J} n_i E_i$ on Star Y. Thus M^Y and hence N^Y are finitely generated free abelian groups, and $M^Y_{\mathbb{R}}, N^Y_{\mathbb{R}}$ are dual finite-dimensional real vector spaces. Moreover M^Y_+ is clearly the intersection of M^Y with a convex rational polyhedral cone in $M^Y_{\mathbb{R}}$ and σ^Y is the dual cone in $N^Y_{\mathbb{R}}$. We can relate the M, N and σ's to our models via:

<u>Lemma 1</u>: <u>Let</u> x <u>be a closed point of</u> Y <u>and</u> (T, X_σ, t) <u>a local model at</u> x. <u>Let</u> E_i <u>correspond to the codimension 1 orbit</u> $\mathcal{O}_i \subset X_\sigma$, <u>and hence to the face</u> $\mathbb{R}^+ \cdot v_i \subset \sigma$. <u>Then the following are equivalent</u>:

i) $\sum n_i E_i$ <u>is a Cartier div. on</u> Star Y

ii) $\sum n_i E_i$ <u>has a local</u> eq^n <u>at</u> x

iii) $\sum n_i \bar{\mathcal{O}}_i$ <u>has a local</u> eq^n <u>at</u> t

iv) $\sum n_i \mathcal{O}_i$ <u>is a Cartier div. on</u> X_σ

v) $\sum n_i \bar{\mathcal{O}}_i = (\mathfrak{X}^r)$, <u>some</u> $r \in M(T)$.

<u>Proof</u>: (ii) $\Longleftrightarrow$ (iii) follows because an ideal $\mathfrak{A}$ is a noetherian local ring $\mathcal{O}$ is principal if and only if $\mathfrak{A} \cdot \hat{\mathcal{O}}$ is principal. Now v) $\Longrightarrow$ iv) $\Longrightarrow$ iii) and i) $\Longrightarrow$ ii) are obvious. Next start with any Weil divisor $D = \sum n_i \bar{\mathcal{O}}_i$. Then $\{y \in X_\sigma \mid D$ has a local eq^n at $y\}$ is an open T-invariant set V. If $t \in V$, it follows that V contains all orbits $\mathcal{O}$ such that $t \in \bar{\mathcal{O}}$, i.e., $V = X_\sigma$. Thus iii) $\Longrightarrow$ iv). iv) $\Longrightarrow$ v) by [I, Th. 8]. Finally to see ii) $\Longrightarrow$ i), it suffices to show, for <u>all</u> closed points $x' \in Y$,

$(\sum n_i E_i$ has a local eq^n at $x') \Longleftrightarrow (\sum n_i E_i$ has a local eq^n at $\eta_Y)$

(η_Y = gen. pt. of Y). Again "$\Longrightarrow$" is obvious. To prove "$\Longleftarrow$", we may as well rename $x' = x$, and use the local model (T, X_σ, t). Let η_0 = gen. pt. of $\mathcal{O}(t) \subset X_\sigma$. Then localizing $\hat{\mathcal{O}}_{X,x} \xrightarrow{\approx} \hat{\mathcal{O}}_{X_\sigma,t}$ with respect to the prime ideals of Y, $\mathcal{O}(t)$ resp., one checks that $\sum n_i E_i$ has a local eq^n at η_Y iff $\sum n_i \bar{\mathcal{O}}_i$ has a local eq^n at η_0. By the same argument as above, $\sum n_i \bar{\mathcal{O}}_i$ has a local eq^n at η_0 implies $\sum n_i \bar{\mathcal{O}}_i$ is a Cartier div. on X_σ, hence $\sum n_i E_i$ has a local eq^n at x. <u>QED</u>

<u>Corollary 1</u>: <u>There are canonical isomorphisms</u>:

i) $M^Y \cong M(T) / \{r \mid r \equiv 0 \text{ on } \sigma\}$

ii) $N^Y_{\mathbb{R}} \cong \{\text{subgp. of } N(T)_{\mathbb{R}} \text{ spanned by } \sigma\}$

iii) $\sigma^Y \cong \sigma$.

<u>Proof</u>: Define the map

$$M(T) \longrightarrow M^Y$$

by

$$r \longmapsto \sum n_i E_i \quad \text{if} \quad (\mathfrak{X}^r) = \sum n_i \bar{\mathcal{O}}_i .$$

By Lemma 1 it is surjective and the kernel is just $\{r \mid (\mathfrak{X}^r) = 0$ or $\mathfrak{X}^r \in \Gamma(\mathcal{O}^*_{X_\sigma})\} = \{r \mid r \equiv 0 \text{ on } \sigma\}$. This implies i) and ii). From the obvious fact that $\sum n_i E_i$ is effective if and only if $\sum n_i \bar{\mathcal{O}}_i$ is effective, it follows that in this map, M^Y_+ = image of $\check{\sigma} \cap M(T)$, hence in the map of (ii), σ^Y corresponds to σ.

Corollary 2: 1) If Z is a stratum in Star Y, then there exists a positive Cartier divisor D on Star Y such that Star Z = Star Y - Supp D.

2) If D is a positive Cartier divisor on Star Y, then Star Y - Supp D = Star Z for some stratum Z ⊂ Star Y.

Proof: 1) Let $\overline{Z}$ correspond formally to $\overline{\mathcal{O}^{\tau}}$ via the isomorphism $\hat{\mathcal{O}}_{X,x} \xrightarrow{\approx} \hat{\mathcal{O}}_{X_\sigma,t}$. There exists $r \in M(T)$ such that $r \geq 0$ on σ, and $\tau = \{\sigma \cap x \mid r(x) = 0\}$. Then $(X_\sigma)_{\mathfrak{X}^r} \cong X_\tau$, i.e., it is the open subset consisting of orbits $\mathcal{O}^{\tau'}$ for all faces τ' of τ. Let $(\mathfrak{X}^r) = \sum n_i \overline{\mathcal{O}}_i$ and let $D = \sum n_i E_i$ be the corresponding divisor in Star Y. Star Y - Supp D is a union of various strata Z', and if $\overline{Z}'$ corresponds formally to $\overline{\mathcal{O}^{\tau'}}$, then

$$
\begin{aligned}
Z' \subset \text{Star } Y - \text{Supp } D &\iff \overline{Z}' \not\subset \text{Supp } D \\
&\iff \overline{\mathcal{O}^{\tau'}} \not\subset \text{Supp}(\mathfrak{X}^r) \\
&\iff \overline{\mathcal{O}^{\tau'}} \supseteq \overline{\mathcal{O}^{\tau}} \\
&\iff \overline{Z'} \supseteq Z \\
&\iff Z' \subseteq \text{Star } Z.
\end{aligned}
$$

2) If $D = \sum n_i E_i$, let $\sum n_i \overline{\mathcal{O}}_i = (\mathfrak{X}^r)$. Since $n_i \geq 0$, $\mathfrak{X}^r \in \Gamma(\mathcal{O}_{X_\sigma})$ and $r \geq 0$ on σ. Let $\tau = \sigma \cap \{x \mid r(x) = 0\}$, and let $\overline{\mathcal{O}^{\tau}}$ correspond formally to $\overline{Z}$ for some stratum $Z \subset$ Star Y. Then the same argument as in (1) shows that Star Y - Supp D = Star Z.

Now we saw in Prop.-Def. 2 that there is a bijection between the strata in Star Y and orbits in X_σ. This now induces further bijections:

$$\{\text{Strata in Star } Y\} \cong \{\text{orbits in } X_\sigma\}$$
$$\cong \{\text{faces of } \sigma\}$$
$$\cong \{\text{faces of } \sigma^Y\}.$$

We can identify this bijection intrinsically on X without use of local models as follows:

I) If $Z_i = E_i - \bigcup_{j \neq i} E_j$ is a codimension 1 stratum in Star Y, and E_i corresponds formally to $\bar{\mathcal{O}}_i \subset X_\sigma$, then $\bar{\mathcal{O}}_i$ corresponds to the ray $\mathbb{R}^+ . v_i \subset N(T)$ where $v_i : M(T) \longrightarrow \mathbb{Z}$ is the linear function:

$$r \longmapsto (\text{order of vanishing of } \mathfrak{X}^r \text{ on } \bar{\mathcal{O}}_i).$$

Therefore the above bijection takes Z_i to the linear function:

$$e_i : \left(\sum_{i \in J} n_i E_i \right) \longmapsto n_i .$$

Note that M_+^Y is, by definition,

$$\left\{ x \in M^Y \middle| e_i(x) \geq 0, \text{ all } i \right\},$$

hence σ^Y is the span of the vectors e_i's.

II) If $Z \subset \text{Star } Y$ is any stratum, then

$$Z = \left[\bigcap_{i\in K} E_i - \bigcup_{i\notin K} E_i\right] \cap \text{Star } Y$$

where $K = \left\{i \middle| Z \subset E_i\right\}$. Then if Z corresponds to τ, for all i, $Z \subseteq E_i \iff \tau \supseteq \mathbb{R}^+ \cdot e_i$. Therefore

$$\tau = \text{the face of } \sigma \text{ spanned by } \{e_i\}_{i\in K} .$$

<u>Definition 4</u>: $\text{R.S.}^U(X) = \left\{ \begin{array}{l} \text{set of } k\text{-morphisms} \\ \lambda: \text{Spec } k[[t]] \longrightarrow X \\ \text{such that } \lambda(\text{gen.pt.}) \in U \end{array} \right\}$

R.S. is short for "Riemann Surface" as used, more or less, by Zariski, [Z-S, vol. II, p.110]. Notice that we have a pairing:

$$\langle , \rangle: \text{ R.S.}^U(\text{Star } Y) \times M^Y \longrightarrow \mathbb{Z}$$

defined by:

$$\langle \lambda, D\rangle = \text{ord}_o(\lambda^{-1}D)$$

where $\quad \text{ord}_o$ = order of vanishing at the closed point 0 of a divisor

This pairing dualizes to a map:

$$\text{ord}: \text{ R.S.}^U(\text{Star } Y) \longrightarrow N^Y$$

and since $\langle \lambda, D\rangle \geq 0$ if D is effective, ord factors through σ^Y:

$$\boxed{\text{ord}: \text{ R.S.}^U(\text{Star } Y) \longrightarrow \sigma^Y \cap N^Y}$$

This is the non-linear analog of the interpretation of $N(T)$ as the 1-P.S.'s of T and of $\sigma \cap N(T)$ as the 1-P.S.'s which extend to morphisms

$\lambda: \mathbb{A}^1_k \longrightarrow X_\sigma$. We will see below that, in fact, $\sigma^Y \cap N^Y$ = Image (ord). For the moment, notice that:

$$(*) \qquad \left(\text{Int } \sigma^Y\right) \cap N^Y \subset \text{Image (ord)}.$$

In fact, taking a closed point $x \in Y$ and a local model (T, X_σ, t), we may assume that t is the distinguished point in its orbit in the sense of [I, Th. 2], i.e., for each $a \in \left(\text{Int}(\sigma)\right) \cap N(T)$, $\lambda_a(o) = t$. Hence λ_a induces $\hat{\lambda}_a$:

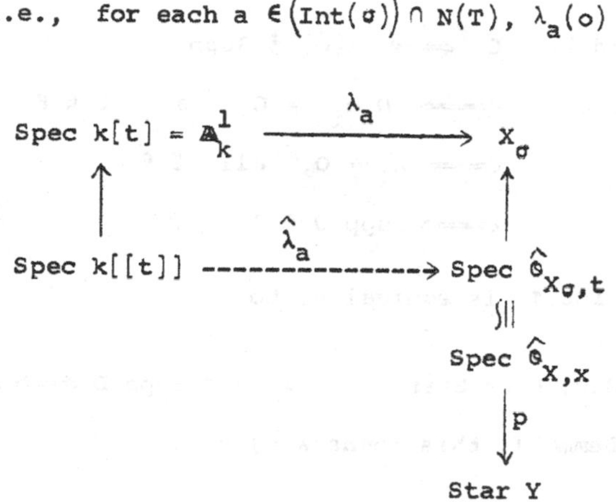

hence $p \cdot \hat{\lambda}_a \in \text{R.S.}^U(\text{Star } Y)$ and it is immediate that in the isomorphism of Corollary to Lemma 1, $a = \text{ord}(p \cdot \hat{\lambda}_a)$.

"ord" also gives us a direct relation between the strata in Star Y and the faces of σ^Y:

<u>Lemma 2</u>: <u>Let</u> $\lambda \in \text{R.S.}^U(\text{Star } Y)$. <u>If</u> $Z \subset$ Star Y <u>is a stratum and</u> τ <u>is the corresponding face of</u> σ^Y <u>as in Prop.-Def.</u> 2, <u>then</u>:

$$\lambda(o) \in Z \iff \text{ord } \lambda \in \text{Int } \tau.$$

Proof: Suppose $K = \left\{ i \middle| Z \subset E_i \right\}$, so that τ = span of $\{e_i\}_{i \in K}$.
Then ord $\lambda \in \text{Int } \tau$ is equivalent to:

$$\forall\, D \in M^Y \text{ such that } D \geq 0 \text{ on } \sigma^Y,$$

$$\langle D, \text{ord } \lambda\rangle = 0 \iff D \equiv 0 \text{ on } \tau.$$

Such a D comes from a positive Cartier divisor $D = \sum n_i E_i$ on Star Y and:

a) $\langle D, \text{ord } \lambda\rangle = 0 \iff \lambda(o) \notin \text{Supp } D$

b) $D \equiv 0$ on $\tau \iff \langle D, e_i\rangle = 0$, all $i \in K$

$\iff n_i = 0$, all $i \in K$

$\iff \text{Supp } D \cap Z = \emptyset$.

Thus ord $\lambda \in \text{Int } \tau$ is equivalent to:

$$\forall \text{ pos.C-div. } D \text{ on Star } Y, \quad \lambda(o) \notin \text{Supp } D \iff Z \cap \text{Supp } D = \emptyset.$$

By Cor. 2 to Lemma 1, this means $\lambda(o) \in Z$. QED

The final step is to glue together all the polyhedral cones σ^Y into one big conical polyhedral complex. We must first compare σ^Y and σ^Z when Z is a stratum in Star Y: There are canonical maps $\alpha^{Y,Z}$ and $\beta^{Y,Z}$ (which we abbreviate to α and β):

a) $M^Y \xrightarrow{\alpha} M^Z$ (restriction of divisors from Star Y to Star Z)

hence

b) $M^Y_{\mathbb{R}} \xrightarrow{\alpha} M^Z_{\mathbb{R}}$

$N^Y \xleftarrow{\beta} N^Z$ (β = dual of α)

$N^Y_{\mathbb{R}} \xleftarrow{\beta} N^Z_{\mathbb{R}}$.

c) $M_+^Y \xrightarrow{\alpha} M_+^Z$ (since res of pos. div. is pos.)

hence

d) $\sigma^Y \xleftarrow{\beta} \sigma^Z$.

<u>Lemma 3</u>: <u>$M^Y \longrightarrow M^Z$ is surjective.</u>

<u>Proof</u>: Let $\{E_i\}_{i \in T}$ be the components of (Star Y)-U. Let $\{E_i\}_{i \in K}$ be those $E_i \subset$ Star Y - Star Z. If D is a Cartier divisor on Star Z - U, we can write

$$D = \sum_{i \in J-K} n_i E_i$$

and extend it by the same formula to a Weil divisor on Star Y. We seek

$$D' = D + \sum_{i \in K} m_i E_i$$

which is a Cartier divisor on Star Y. If $x \in Y$, it suffices by Lemma 1 that D' have a local equation at x. Let (X_σ, t) be a local model at x and let

$$D^* = \sum_{i \in J-K} n_i \bar{\mathcal{O}}_i$$

be the corresponding Weil divisor. If Z corresponds to the face τ of σ, then by the argument used in Lemma 1,

$$D \text{ Cartier div. in Star } Z \Longrightarrow D^* \text{ Cartier div. in } X_\tau.$$

But then by Lemma 1, $D^*|_{X_\tau} = (\mathfrak{X}^r)|_{X_\tau}$ for some $r \in M(T)$. Then let $D^{*\prime} = (\mathfrak{X}^r)$: this is a Cartier div. on X_σ extending $D^*|_{X_\tau}$. Let D' be the corresponding divisor on X. <u>QED</u>

Corollary 1: $N^Z_{\mathbb{R}} \longrightarrow N^Y_{\mathbb{R}}$ is injective and considering this as an inclusion: $N^Z = N^Z_{\mathbb{R}} \cap N^Y$.

Corollary 2: If Z corresponds to the face τ of σ^Y, then the inclusion $N^Z_{\mathbb{R}} \longrightarrow N^Y_{\mathbb{R}}$ maps σ^Z isomorphically onto τ.

Proof: Using the notation of the lemma, let $e_i \in N^Y$ be the map $\sum n_j E_j \longmapsto n_i$, $i \in J$. Then if $i \in J-K$, $e_i(\sum n_j E_j)$ depends only on the restriction of $\sum n_j E_j$ to Star Z, hence these e_i are in N^Z. Now we know that since Star Z-U $= \bigcup_{i \in J-K} E_i$, σ^Z is spanned by $\{e_i\}_{i \in J-K}$. On the other hand, in $M^Y_{\mathbb{R}}$, we know that τ is spanned by $\{e_i\}_{i \in J-K}$. Therefore σ^Z goes onto τ in the inclusion $M^Z_{\mathbb{R}} \hookrightarrow M^Y_{\mathbb{R}}$.

Corollary 3: The diagram

$$\begin{array}{ccc} \text{R.S.}^U(\text{Star } Z) & \xrightarrow{\text{ord}} & \sigma^Z \\ \cap & & \downarrow \beta \\ \text{R.S.}^U(\text{Star } Y) & \xrightarrow{\text{ord}} & \sigma^Y \end{array}$$

commutes, hence

$$\text{ord}: \ \text{R.S.}^U(\text{Star } Y) \longrightarrow \sigma^Y \cap N^Y$$

is surjective.

Proof: The commutativity is immediate from the definitions and then surjectivity of ord follows from the fact that for all Z, $\text{Im}(\text{ord}^{(Z)}) \supset (\text{Int } \sigma^Z) \cap N^Z$, hence by Cor. 1:

$$\text{Im}(\text{ord}^{(Y)}) \supset (\text{Int of face } \tau \text{ corresp. to } Z) \cap N^Y.$$

Next two general definitions on the kind of object we are seeking to define

Definition 5: A $\left\{\begin{matrix}\text{conical}\\ \text{compact}\end{matrix}\right\}$ polyhedral complex Δ is a topological space $|\Delta|$ plus a finite family of closed subsets $\sigma_\alpha \subset |\Delta|$ called its cells plus a finite-dimensional real vector space V_α of real-valued continuous functions on σ_α such that

1(conical case): via a basis $f_1,\cdots,f_{n_\alpha}$ of V_α, we get a homeomorphism

$$\phi_\alpha : \sigma_\alpha \xrightarrow{\approx} \sigma'_\alpha \subset \mathbb{R}^{n_\alpha},$$

where σ'_α is a conical convex polyhedra in $\mathbb{R}^{n_\alpha}$, not contained in a hyperplane,

1 (compact case): $V_\alpha \supset \mathbb{R}$, the constant functions, and via a basis $1, f_1,\cdots,f_{n_\alpha}$ of V_α, we get a homeomorphism:

$$\phi_\alpha : \sigma_\alpha \xrightarrow{\approx} \sigma'_\alpha \subset \mathbb{R}^{n_\alpha}$$

where σ'_α is a compact convex polyhedron in $\mathbb{R}^{n_\alpha}$, not contained in a hyperplane,

2: ϕ_α^{-1} (faces of σ'_α) = other σ_β's, which we call the faces of σ_α; we call $\phi_\alpha^{-1}(\text{Int } \sigma'_\alpha)$ the interior of σ_α

3: $|\Delta| = \{\text{disj. union of Int } \sigma_\alpha, \text{ all } \alpha\}$

4: if $\sigma_\beta = (\text{a face of } \sigma_\alpha)$, then $\text{res}_{\sigma_\beta} V_\alpha = V_\beta$.

We can "explain" the idea of V_α as follows: we want σ_α to be a conical or compact polyhedron in an actual real vector space, but unique only up to linear transformations, or affine transformations in the 2 cases.

__Definition 6: An integral structure on a $\left\{\begin{matrix}\text{conical}\\ \text{compact}\end{matrix}\right\}$ polyhedral complex is a set of finitely generated abelian groups $L_\alpha \subset V_\alpha$ such that:__

1 (compact case only): $L_\alpha \supset n\mathbb{Z}$, the constant functions with values in $n\mathbb{Z}$, for some n,

2: $L_\alpha \otimes \mathbb{R} \xrightarrow{\approx} V_\alpha$,

3: If σ_β is a face of σ_α, then $\mathrm{res}_{\sigma_\beta} L_\alpha = L_\beta$.

__Remark:__ If Δ is a compact polyhedral complex, we can "expand" Δ to a conical polyhedral complex Δ' canonically:

$$|\Delta'| = |\Delta| \times [0,\infty) \Big/ \begin{matrix}\text{identify } |\Delta|\times(0)\\ \text{to one point}\end{matrix}$$

setting

$$\sigma'_\alpha = \sigma_\alpha \times [0,\infty) \Big/ \begin{matrix}\text{identify } \sigma_\alpha\times(0)\\ \text{to one point}\end{matrix}$$

$$V'_\alpha = \text{functions on } \sigma'_\alpha \text{ of form } (x,t) \longmapsto t.f(x),\ f \in V_\alpha.$$

Given a conical polyhedral complex Δ, if f is a continuous function on $|\Delta|$ such that

a) $\mathrm{res}_{\sigma_\alpha} f \in V_\alpha$, all α,

b) $f(x) \geq 0$, all x with equality only if x = apex

then we get a compact polyhedral complex Δ_0:

$$|\Delta_0| = \left\{x \in |\Delta| \,\middle|\, f(x) = 1\right\}$$

$$(\sigma_\alpha)_0 = \sigma_\alpha \cap |\Delta_0|$$

$$(V_\alpha)_0 = \mathrm{res}_{(\sigma_\alpha)_0} V_\alpha .$$

These compact fellows will be quite handy in §4.

Our final point is simply:

<u>To every toroidal embedding without self-intersection</u> $U \subset X$, <u>we can associate a conical polyhedral complex with integral structure</u> $\Delta = (|\Delta|, \sigma^Y, M^Y)$ <u>whose cells are in 1-1 correspondence with the strata of</u> X.

This is now quite simple to define. We set

$$|\Delta| = \coprod_{\text{strata } Y} \sigma^Y \Big/ \left\{ \begin{array}{l} \text{modulo equivalence relation} \\ \text{generated by isomorphisms } \beta^{Y,Z}: \sigma^Z \xrightarrow{\approx} \left(\begin{array}{c}\text{face}\\ \text{of } \sigma^Y\end{array}\right) \\ \text{whenever } Z \subset \text{Star } Y \end{array} \right\}$$

Explicitly, the equivalence relation is:

$$x_1 \in \sigma^{Y_1} \text{ eq. to } x_2 \in \sigma^{Y_2} \iff \left\{ \begin{array}{l} \text{the faces } \tau_i \text{ of } \sigma^{Y_i} \text{ containing } x_i \\ \text{correspond to the same stratum} \\ Z \subset \text{Star } Y_1 \cap \text{Star } Y_2 \text{ and } x_1, x_2 \\ \text{correspond to the same point of } \sigma^Z \end{array} \right\}$$

so that $|\Delta| = \left\{ \text{disj. union of Int } \sigma^Y \right\}$ and the identification is carried out like this: $\forall\, Y_1, Y_2$, $\text{Star } Y_1 \cap \text{Star } Y_2$ is an open set in $\text{Star } Y_1$ and a union of strata; hence it is the set of strata corresponding to the set of faces of a closed subpolyhedron $\sigma^{Y_1,Y_2} \subset \sigma^{Y_1}$. Define

$$\begin{array}{ccc} \sigma^{Y_1,Y_2} & \xrightarrow[\approx]{h^{Y_1,Y_2}} & \sigma^{Y_2,Y_1} \\ \cap & & \cap \\ \sigma^{Y_1} & & \sigma^{Y_2} \end{array}$$

by requiring that for all $Z \subset \text{Star } Y_1 \cap \text{Star } Y_2$, h^{Y_1,Y_2} equals $\beta^{Y_2,Z} \cdot (\beta^{Y_1,Z})^{-1}$:

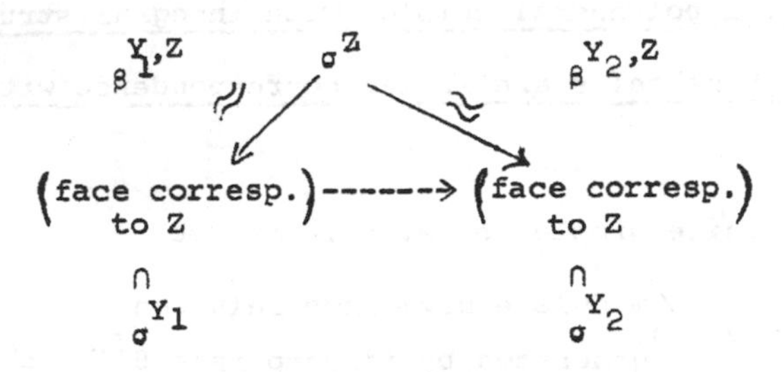

on the face corresponding to Z. For this to be possible, there is a compatibility condition to check whenever $W \subset \text{Star } Z$, $Z \subset \text{Star } Y$, i.e., that the diagram

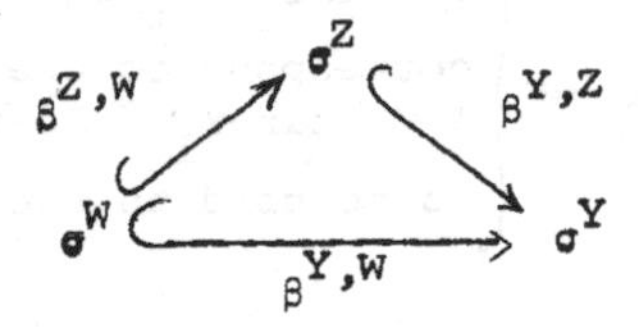

is commutative. This is immediate.

<u>Remark</u>: It also follows immediately that all the "ord" maps patch together to one big "ord" map:

$$\begin{array}{ccc} \text{R.S.}^U(\text{Star } Y) & \xrightarrow{\text{ord}} & \sigma^Y \\ \cap & & \downarrow \\ \text{R.S.}^U(X) & \xrightarrow{\text{ord}} & |\Delta| \end{array} .$$

§2. Toroidal embeddings-theorems

We should like to study to what extent a toroidal embedding is determined by Δ. More precisely, one cannot, of course, expect that Δ determines $U \subset X$; rather if you fix one toroidal embedding $U \subset X$, it turns out that polyhedral subdivisions of the associated Δ determine canonically new toroidal embeddings $U \subset Z$ dominating $U \subset X$:

$$\begin{array}{ccc} & & Z \\ & \nearrow & \downarrow \text{ birational morphism} \\ U & \searrow & X \end{array}$$

We work up to this in a sequence of theorems that we number analogously to those in Ch. I, §1-2. For the whole of this section, fix a particular toroidal embedding $U \subset X$ without self-intersection. The following idea is due to Hironaka:

Definition 1: A birational morphism f:

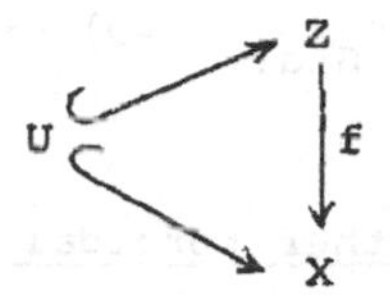

is called canonical if for all $x_1, x_2 \in X$ in the same stratum Y and all

$$\alpha: \hat{\mathcal{O}}_{X,x_1} \xrightarrow{\approx} \hat{\mathcal{O}}_{X,x_2}$$

which preserve the strata, i.e., if $Y \subset \overline{Y^*}$ for some stratum Y^*, then α takes the ideal of $\overline{Y^*}$ at x_1 to the ideal of $\overline{Y^*}$ at x_2, α lifts:

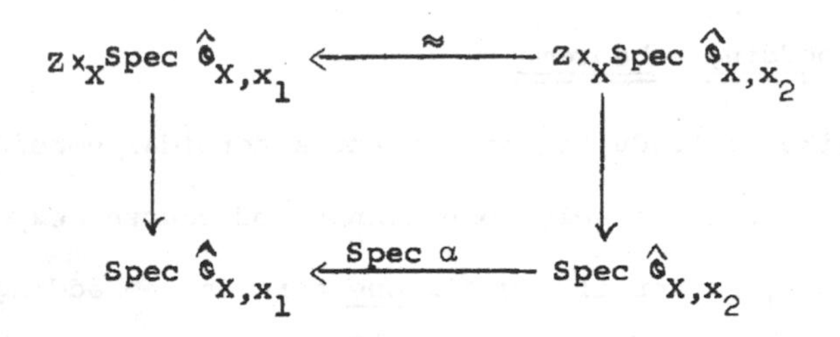

Now fix a stratum $Y \subset X$ and consider diagrams:

$$(*) \qquad \begin{array}{ccc} & & Z \\ & \nearrow & \big\downarrow f \\ U & & \\ & \searrow & \\ & & \text{Star } Y \end{array}$$

f affine and canonical

Z normal

Theorem 1*. There is a 1-1 correspondence between the set of diagrams (*) and the set of rational polyhedral cones $\tau \subset \sigma^Y$ given by

$$\tau \longmapsto Z_\tau = \underline{\mathrm{Spec}}\ \mathfrak{U}_\tau$$

where $\mathfrak{U}_\tau$ = subsheaf $\sum_{D \in \check{\tau} \cap M^Y} \mathcal{O}_{\mathrm{Star}\ Y}(-D)$ of $\underline{\mathbb{R}(X)}$.

For all such diagrams, $U \subset Z$ is another toroidal embedding without self-intersection, and with a unique closed stratum $\tilde{Y}$. Moreover, $\exists$ unique linear isomorphism γ making the following commute:

$$\begin{array}{ccc} R.S.^U(Z) & \xrightarrow{\text{ord}} & \sigma^{\tilde{Y}} \overset{\sim}{\dashrightarrow} \tau \\ \cap & & \\ R.S.^U(\text{Star } Y) & \xrightarrow{\text{ord}} & \sigma^Y \supset \end{array}$$

<u>and if</u> $\lambda \in R.S.^U(\text{Star } Y)$, <u>then</u>

$$\lambda \in R.S.^U(Z) \Longleftrightarrow \text{ord } \lambda \in \tau.$$

<u>Proof</u>: To make Z_τ more explicit, note that we can construct it as follows: let $D_1,\cdots,D_N$ be a basis of the semi-group $\check{\tau} \cap M^Y$. Then if $V \subset \text{Star } Y$ is any open set where each D_i has a local equation δ_i, then

$$\mathfrak{A}_\tau\big|_V = \mathcal{O}_V[\delta_1,\cdots,\delta_N].$$

In particular, $\mathfrak{A}_\tau$ is quasi-coherent and Z_τ is well-defined. It is evident that $f: Z_\tau \longrightarrow \text{Star } Y$ is affine and canonical.

On the other hand, abbreviating σ^Y by σ, we can use the theory of Ch. I, §1, to construct affine embeddings of an n-dimensional torus T:

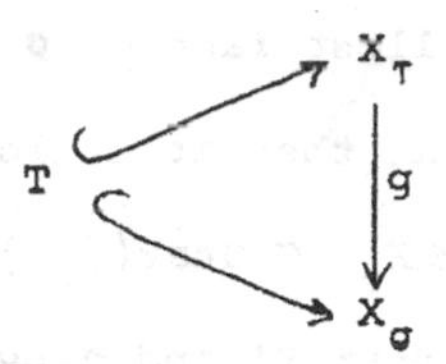

For any closed point $x \in \text{Star } Y$, one can choose a suitable orbit in X_σ, a closed point t in this orbit and a local model $\hat{\mathcal{O}}_{X,x} \cong \hat{\mathcal{O}}_{X_\sigma,t}$. In this isomorphism, δ_i corresponds to $u_i \mathfrak{X}^{r_i}$ for some unit u_i, and

$r_i \in M(T)$. Then $r_1,\cdots,r_N$ generate the corresponding semi-group $\check{\tau} \cap M(T)$ in $M(T)$, and $X_\tau = \underset{\sim\sim}{\text{Spec}}\, \mathcal{O}_{X_\sigma}[\mathfrak{X}^{r_1},\cdots,\mathfrak{X}^{r_N}]$. Therefore our formal isomorphism lifts to ϕ:

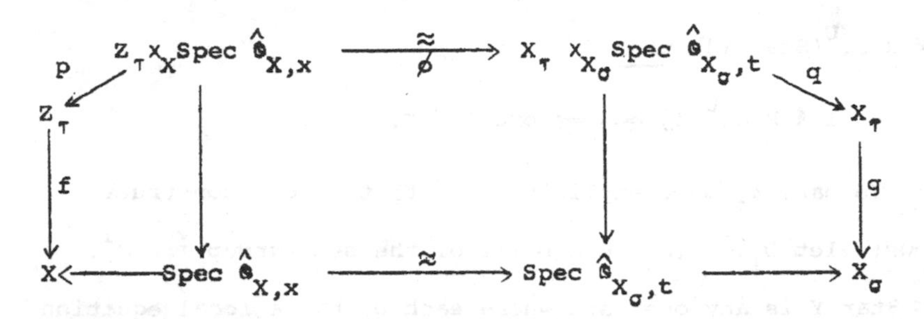

Passing to completions at a point $x' \in Z_\tau$ over x and the corresponding point $t' \in X_\tau$, this gives us simultaneously local models at all points of Z_τ over x, hence proves that Z_τ is normal and that $U \subset Z_\tau$ is a toroidal embedding. Moreover since in the morphism $X_\tau \longrightarrow X_\sigma$, each orbit of X_τ is smooth over some orbit of X_σ, it follows that every stratum of Z_τ is smooth over some stratum of Star Y.

We may make a simplifying reduction to the case $\tau \cap \text{Int}\, \sigma \neq \phi$ at this point. Because if σ' = smallest face of σ containing τ and σ' corresponds to the stratum $Y' \subset X$, then it follows from the existence of ϕ and the fact that $\text{Im}(X_\tau \longrightarrow X_\sigma) \subset \text{Star}(\mathcal{O}_{\sigma'})$ that $\text{Im}(Z_\tau \longrightarrow X) \subset$ Star Y'. So let's replace X by Star Y' and σ by σ'.

To go further we need:

<u>Lemma 1</u>: <u>For all orbits</u> $\mathcal{O} \subset X_\tau$, $q^{-1}(\bar{\mathcal{O}})$ <u>is irreducible and normal.</u>

<u>Proof</u>: Note that q is a flat morphism with regular fibres, hence $\bar{\Phi}$ normal $\Longrightarrow q^{-1}(\bar{\Phi})$ normal (EGA IV.6.5.4). If Φ_o = image of Φ in X_σ, note that

a) $\bar{\Phi}_o$ is normal

b) Since the stabilizer of a point of Φ_o is connected, $\Phi \cong \Phi_o \times T'$ for some torus T', hence $\mathbb{R}(\Phi_o)$ is algebraically closed in $\mathbb{R}(\Phi)$.

Irreducibility now follows from:

<u>Lemma 2</u>: <u>Let</u> $f: X \longrightarrow Y$ <u>be a morphism of varieties</u>, $Y_1 = \overline{f(X)}$ <u>and let</u> $y \in Y_1$. <u>If</u>

a) Y_1 <u>is normal at</u> y

b) $\mathbb{R}(Y_1)$ <u>is algebraically closed in</u> $\mathbb{R}(X)$, <u>then</u>:

$$X \times_Y \operatorname{Spec} \hat{\mathcal{O}}_{y,Y} \text{ is irreducible.}$$

<u>Proof</u>: Since $X \times_Y \operatorname{Spec} \hat{\mathcal{O}}_{y,Y}$ is flat over X, all its generic points lie over the generic point of X. Thus it is enough to show that $\mathbb{R}(X) \otimes_{\mathcal{O}_{y,Y}} \hat{\mathcal{O}}_{y,Y}$ is an integral domain. But

$$\mathbb{R}(X) \otimes_{\mathcal{O}_{y,Y}} \hat{\mathcal{O}}_{y,Y} \cong \mathbb{R}(X) \otimes_{\mathbb{R}(Y_1)} \left[\mathbb{R}(Y_1) \otimes_{\mathcal{O}_{y,Y_1}} \hat{\mathcal{O}}_{y,Y_1}\right].$$

By (a), $\hat{\mathcal{O}}_{y,Y_1}$ is a domain and $\mathbb{R}(Y_1) \otimes_{\mathcal{O}_{y,Y_1}} \hat{\mathcal{O}}_{y,Y_1}$ is part of its quotient field, which is separable over $\mathbb{R}(Y_1)$. Therefore by (b), the whole thing is still a domain. <u>QED</u>

Now assume that we have chosen a point $x \in Y$, so that the corresponding point t is in the closed orbit of X_σ. In this case:

a) $\forall$ strata $W \subset Z_\tau$, $\eta_W \in$ Image p

b) $\forall$ orbits $\mathcal{O} \subset X_\tau$, $\eta_{\mathcal{O}} \in$ Image q

because the image of η_W in X (resp. of $\eta_{\mathcal{O}}$ in X_σ) equals the generic point of some stratum (resp. some orbit) which lies in Spec $\mathcal{O}_{x,X}$ (resp. Spec $\mathcal{O}_{t,X_\sigma}$). Let

$$E_1, \cdots, E_N = \text{comp. of } Z_\tau - U$$
$$\overline{\mathcal{O}}_1, \cdots, \overline{\mathcal{O}}_M = \text{comp. of } X_\tau - T.$$

It follows from Lemma 1 that the sets:

1. of schemes $q^{-1}(\overline{\mathcal{O}}_i)$
2. of components of schemes $q^{-1}(\overline{\mathcal{O}}_i)$
3. of components of $X_\tau \times_{X_\sigma} \text{Spec } \hat{\mathcal{O}}_t - q^{-1}(T)$
4. of components of $Z_\tau \times_X \text{Spec } \hat{\mathcal{O}}_x - p^{-1}(U)$
5. of components of schemes $p^{-1}(E_i)$

are all equal or in 1-1 correspondence. We would like to show that the schemes $p^{-1}(E_i)$ are irreducible so that we can add to our list the set:

6. of schemes $p^{-1}(E_i)$.

Suppose to the contrary $p^{-1}(E_{i_o})$ has ≥ 2 components which correspond say to $q^{-1}(\overline{\mathcal{O}}_{i_1})$ and $q^{-1}(\overline{\mathcal{O}}_{i_2})$. We can find a character $\mathfrak{X}^r$ of T with $r \geq 0$ on τ such that

a) $\chi^r \equiv 0$ on $\overline{\mathcal{O}}_{i_1}$

b) χ^r unit generically on $\overline{\mathcal{O}}_{i_2}$.

But (χ^r) corresponds formally to a Cartier divisor D on Z_τ supported on $Z_\tau - U$, and $r \geq 0$ on τ implies D effective and a) and b) imply:

a') $D \equiv 0$ on the branch of E_{i_o} corresponding to $\overline{\mathcal{O}}_{i_1}$

b') $D \not\equiv 0$ on the branch of E_{i_o} corresponding to $\overline{\mathcal{O}}_{i_2}$.

But E_{i_o} is irreducible so D has a definite multiplicity along E_{i_o} which is either positive or zero, i.e., a' and b' are incompatible. Thus $p^{-1}(E_{i_o})$ is irreducible.

This shows incidentally that for each i, $\phi(p^{-1}(E_i)) = q^{-1}(\overline{\mathcal{O}}_j)$ for some j, hence the E_i are normal. Therefore $U \subset Z_\tau$ is a toroidal embedding without self-intersection. But more than that, we can show that Z_τ has a unique closed stratum. In fact, let $\mathcal{O}^*$ be the closed orbit in X_τ and let $t^* \in \mathcal{O}^*$ be a closed point lying over $t \in X_\sigma$. This gives us $(t^*,t) \in X_\tau \times_{X_\sigma} \operatorname{Spec} \hat{\mathcal{O}}_{t,X_\sigma}$ and $x^* = p(\phi^{-1}(t^*,t)) \in Z_\tau$. Then I claim:

> $\forall$ strata $W \subset Z_\tau$, $x^* \in \overline{W}$
>
> hence the stratum $\widetilde{Y}$ containing x^* is the only closed stratum.

In fact, suppose $\overline{W}$ is a component of $\bigcap_{i \in J} E_i$. Then $p^{-1}(\overline{W})$, even if it does not remain irreducible, still contains a component of $\bigcap_{i \in J} p^{-1}(E_i)$.

Hence for some J'

$$\phi(p^{-1}(\overline{W})) \supseteq \text{a component of } \bigcap_{i\in J'} q^{-1}(\overline{\mathcal{O}}_i) = q^{-1}(\bigcap_{i\in J'} \overline{\mathcal{O}}_i) .$$

But $\bigcap_{i\in J'} \overline{\mathcal{O}}_i$ is, in fact, the closure of an orbit $\overline{\mathcal{O}}$ (i.e., let τ' = least face of τ containing the vertices in J': then let $\mathcal{O}$ correspond to τ'). And we have seen that $q^{-1}(\overline{\mathcal{O}})$ is irreducible. Therefore

$$\phi(p^{-1}(\overline{W})) \supseteq q^{-1}(\overline{\mathcal{O}}).$$

Since $\mathcal{O}^* \subseteq \overline{\mathcal{O}}$, $t^* \in q^{-1}(\overline{\mathcal{O}})$. Therefore $x^* \in \overline{W}$.

This proves that Z_τ = Star $\widetilde{Y}$. Finally, to compare τ and $\sigma^{\widetilde{Y}}$, pass to the completion of $Z_\tau \times_{X_\sigma} \mathrm{Spec}\, \hat{\mathcal{O}}_{t,X_\sigma}$ at (t^*,t) to get local models for X and Z_τ simultaneously:

$$\begin{array}{ccc} \hat{\mathcal{O}}_{Z_\tau,x^*} & \cong & \hat{\mathcal{O}}_{X_\tau,t^*} \\ \uparrow & & \uparrow \\ \hat{\mathcal{O}}_{X,x} & \cong & \hat{\mathcal{O}}_{X_\sigma,t} \end{array} .$$

As above let E_i and $\overline{\mathcal{O}}_i$ be corresponding components of $Z_\tau - U$ and $X_\tau - T$. Then as in §1:

$$\begin{aligned} \sum n_i E_i \text{ is a Cartier div. in } Z_\tau &\Longrightarrow \sum n_i \overline{\mathcal{O}}_i \text{ is a Cartier div. in } X_\tau \\ &\Longrightarrow \sum n_i \overline{\mathcal{O}}_i = (\mathfrak{X}^r), \text{ some } r \in M(T) \\ &\Longrightarrow \sum n_i \overline{\mathcal{O}}_i = g^*D, \text{ D a Cartier div. on } X_\sigma \end{aligned}$$

$\Longrightarrow \sum n_i E_i$ agrees formally at x^* with f^*D, D a Cartier divisor on X

$\Longrightarrow \sum n_i E_i = f^*D$, D a Cartier div. on X.

In other words, $D \longmapsto f^*D$ sets up

$$(*) \qquad M^Y \cong M^{\tilde{Y}},$$

hence

$$(**) \qquad N_{\mathbb{R}}^Y \cong N_{\mathbb{R}}^{\tilde{Y}}.$$

But also:

$\sum n_i E_i$ effective $\Longleftrightarrow \sum n_i \overline{\mathcal{O}}_i$ effective

$\Longleftrightarrow$ the corresp. linear function $\ell_{\sum n_i \overline{\mathcal{O}}_i}$ is ≥ 0 on τ,

hence in the identification (*)

$$M_+^{\tilde{Y}} \cong \check{\tau} \cap M^{\tilde{Y}}$$

and in (**)

$$\sigma^{\tilde{Y}} \cong \tau.$$

It is immediate that isomorphism is compatible with ord since for all λ: Spec $k[[t]] \longrightarrow Z_\tau$, and $D \in M^Y$,

$$\langle f\cdot\lambda, D\rangle = \langle\lambda, f^*D\rangle.$$

And if λ: Spec $k[[t]] \longrightarrow X$ is given, then clearly λ lifts to Z_τ if and only if

$\lambda^*(\delta_i) \in k[[t]]$, ($\delta_i$ = local eq^n of D_i, D_i's generators of $\check{\tau} \cap M^Y$)

ie.,

$$\langle \text{ord } \lambda, D_i \rangle \geq 0, \qquad 1 \leq i \leq N,$$

and this is equivalent to ord $\lambda \in \check{\check{\tau}} = \tau$.

It remains to prove that all affine canonical modifications of Star Y are obtained in this way. Let

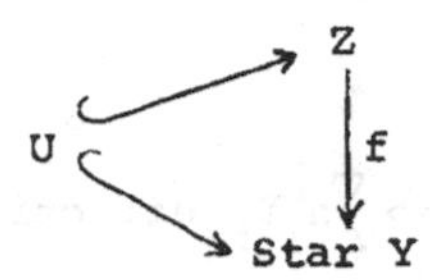

be a canonical affine modification. For all $D \in M_+^Y$, define a sheaf of ideals:

$$\mathcal{J}_D = \left\{ a \in \mathcal{O}_X \,\middle|\, \frac{a}{\delta} \in f_*\mathcal{O}_Z, \quad \delta = \text{local equation of } D \right\}.$$

Then because $f_*\mathcal{O}_Z|_U = \mathcal{O}_{\text{Star } Y}|_U$ and $f_*\mathcal{O}_Z$ is quasi-coherent, it follows easily that

$$f_*\mathcal{O}_Z = \bigcup_{D \in M_+^Y} \mathcal{J}_D \cdot \mathcal{O}_{\text{Star } Y}(D).$$

Moreover, because f is canonical, so is $\mathcal{J}_D$ (i.e., $\mathcal{J}_D$ is invariant under all formal isomorphisms α as in Def. 1). In this way, we can easily reduce the proof that $Z \cong Z_\tau$, some τ, to:

<u>Lemma 3: Let</u> $\mathfrak{J}$ <u>be a canonical coherent sheaf of fractional ideals on</u> Star Y. <u>Then</u> $\exists\ D_1,\cdots,D_n \in M^Y$ <u>such that</u>

$$\mathfrak{J} = \sum_{i=1}^{n} \mathfrak{O}_{\text{Star } Y}(D_i).$$

<u>Proof</u>: Let $\{z_1,\cdots,z_d\} = \operatorname{Ass}(\mathfrak{J})$ and $Z_i = \overline{\{z_i\}}$. Let

$$x \in Y - \bigcup_{\substack{\text{all } i \text{ s.t.} \\ Z_i \not\supseteq Y}} Z_i$$

be a closed point and $\phi\colon \hat{\mathfrak{O}}_{x,X} \cong \hat{\mathfrak{O}}_{t,X_\sigma}$ a local model. We prove the lemma in 4 steps:

Step I: $\exists\ r_1,\cdots,r_n \in M(T)$ s.t. $\phi(\mathfrak{J}\cdot\hat{\mathfrak{O}}_{x,X}) = (\mathfrak{X}^{r_1},\cdots,\mathfrak{X}^{r_n})$,

Step II: $\exists\ D_1,\cdots,D_n \in M^Y$ s.t. $\mathfrak{J}_x = \sum \mathfrak{O}_{\text{Star } Y}(D_i)_x$

Step III: $\exists\ D_1,\cdots,D_n \in M^Y$ and a neighborhood V of x s.t.

$$\mathfrak{J}|_V = \sum \mathfrak{O}_{\text{Star } Y}(D_i)|_V$$

Step IV: $\exists\ D_1,\cdots,D_n \in M^Y$ s.t. $\mathfrak{J} = \sum \mathfrak{O}_{\text{Star } Y}(D_i)$.

<u>Step I</u>: The fractional ideal $\mathfrak{G} = \phi(\mathfrak{J}\cdot\hat{\mathfrak{O}}_{x,X})$ in $\hat{\mathfrak{O}}_{t,X_\sigma}$ is by hypothesis invariant under all automorphisms of $\hat{\mathfrak{O}}_{t,X_\sigma}$ that leave fixed the components of $X_\sigma - T$. Let $\mathfrak{O} \subset X_\sigma$ be the orbit through t which we can assume closed. Then also for all associated primes $\mathfrak{p}$ of $\mathfrak{G}$, $\mathfrak{p} \subseteq \mathfrak{p}' = I(\mathfrak{O})\cdot\hat{\mathfrak{O}}_t$. If T_1 = stabilizer of t, then we can write

$$T = T_1 \times T_2$$
$$X_\sigma = X_1 \times T_2$$
$$\Phi = \{t_1\} \times T_2$$
$$t = (t_1, 1).$$

Now embed $T_2 \cong (\mathbb{A}^1 - (o))^k \subset \mathbb{A}^k$ in the usual way, and consider

$$T \subset X'_\sigma = X_1 \times \mathbb{A}^k .$$

Define

$$\psi: X_\sigma \longrightarrow X'_\sigma$$
$$\psi(x_1, t_2) = (x_1, t_2 - 1)$$

taking the point t to the point $t' = (t,0) \in X'_\sigma$. Via ψ, now t' is a <u>fixed point</u> for T acting on X'_σ, so T acts on $\mathcal{O}_{t',X'_\sigma}$, hence <u>via ψ</u> on $\mathcal{O}_{t,X_\sigma}$ and $\hat{\mathcal{O}}_{t,X_\sigma}$ as well. Therefore by assumption $\mathcal{G}$ is invariant under this action. By the usual argument this means that $\mathcal{G}$ is generated by characters for this action, i.e., by $\mathfrak{X}^r \cdot \psi$, $r \in M(T)$. But now $\mathcal{G} = \mathcal{G} \cdot (\hat{\mathcal{O}}_t)_{\mathfrak{p}'} \cap \hat{\mathcal{O}}_t$ and in $(\hat{\mathcal{O}}_t)_{\mathfrak{p}'}$, all the characters $\mathfrak{X}^r \cdot \psi$, $r \in M(T_2)$, become units. But $\mathfrak{X}^r \cdot \psi = \mathfrak{X}^r$ if $r \in M(T_1)$, so $\mathcal{G}$ is generated by characters $\mathfrak{X}^r$.

<u>Step I $\Longrightarrow$ Step II</u> since if δ_i is a local equation of E_i, then $\phi(\delta_i) = u_i \cdot \mathfrak{X}^{r_i}$, u_i unit and r_i a basis of $\check{\sigma} \cap M(T)$.

<u>Step II $\Longrightarrow$ Step III</u> by coherency.

Step III $\Longrightarrow$ Step IV: In fact, for all closed points $x' \in \text{Star } Y$, choose $x'' \in V$ in the same stratum as x'. Then there exists an isomorphism

$$\psi\colon \hat{\mathcal{O}}_{x',X} \cong \hat{\mathcal{O}}_{x'',X}$$

preserving strata. Since $\mathfrak{J}$ is canonical, we get:

$$\begin{aligned}\psi(\mathfrak{J}\cdot\hat{\mathcal{O}}_{x',X}) &= \mathfrak{J}\cdot\hat{\mathcal{O}}_{x'',X} \\ &= \sum \hat{\mathcal{O}}_{x'',X}(D_i) \\ &= \psi(\sum \hat{\mathcal{O}}_{x',X}(D_i))\end{aligned}$$

and hence:

$$\mathfrak{J}_{x'} = \mathcal{O}_{x',X} \cap \mathfrak{J}\cdot\hat{\mathcal{O}}_{x',X} = \mathcal{O}_{x',X} \cap \sum \hat{\mathcal{O}}_{x',X}(D_i) = \sum \mathcal{O}_{x',X}(D_i).$$

QED

Having proven Th. 1*, we can now make rapid progress! The analog of Th. 2, Ch. I has already been worked out in §1 in the very definition of σ^Y and ord. We get next:

Theorem 3*: If $\tau_1, \tau_2 \subset \sigma^Y$ are rational polyhedral cones, then there exists a morphism g:

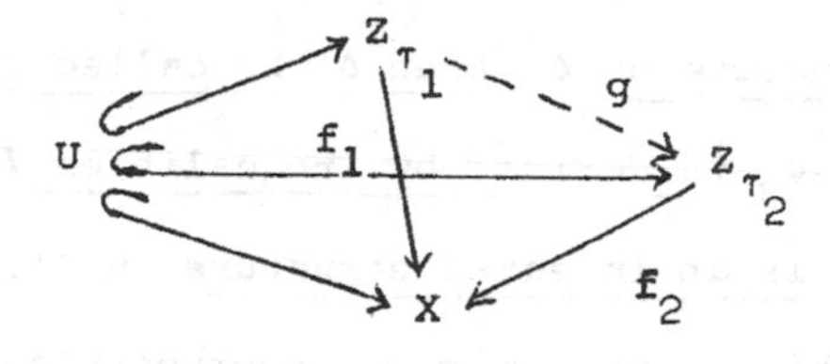

if and only if $\tau_1 \subseteq \tau_2$. Moreover g is an open immersion if and only if τ_1 is a face of τ_2.

The proof is completely analogous to that of Th. 2 replacing orbits by strata, 1-P.S. by elements of $R.S.^U(X)$, and descriptions of the affine rings of X_{τ_1}, X_{τ_2} by descriptions of the sheaves

$f_{1,*}\mathcal{O}_{Z_{\tau_1}}$, $f_{2,*}\mathcal{O}_{Z_{\tau_2}}$.

Theorem 4*: Z_τ is non-singular if and only if $\check{\tau} \cap N^Y$ can be generated by a subset of a basis of N^Y over $\mathbf{Z}$.

The proof this time uses the known result Th. 4 and carries it over to Z_τ by using local models $\hat{\mathcal{O}}_{x,Z_\tau} \cong \hat{\mathcal{O}}_{t,X_\tau}$.

Definition 2: If $\Delta = \left\{|\Delta|, \sigma_\alpha, V_\alpha\right\}$ is a conical polyhedral complex, then a finite partial polyhedral decomposition is a second conical polyhedral complex $\Delta' = \left\{|\Delta'|, \sigma'_\beta, V'_\beta\right\}$ with

1. $|\Delta'| \subseteq |\Delta|$
2. $\forall \beta$, $\exists \alpha$ s.t. $\operatorname{Int} \sigma'_\beta \subseteq \operatorname{Int} \sigma_\alpha$
3. If $\sigma'_\beta \subset \sigma_\alpha$, then $V'_\beta = \operatorname{res}_{\sigma'_\beta} V_\alpha$.

If $\{L_\alpha\}$ is an integral structure on Δ, then Δ' is called rational if whenever $\sigma'_\beta \subseteq \sigma_\alpha$, then σ'_β is defined by inequalities $\ell \geq 0$, $\ell \in L_\alpha$. In this case, $L'_\beta = \operatorname{res}_{\sigma'_\beta} L_\alpha$ is an integral structure on Δ'. As in Chapter I, we abbreviate this to an f.r.p.p. decomposition.

Definition 3: Consider diagrams:

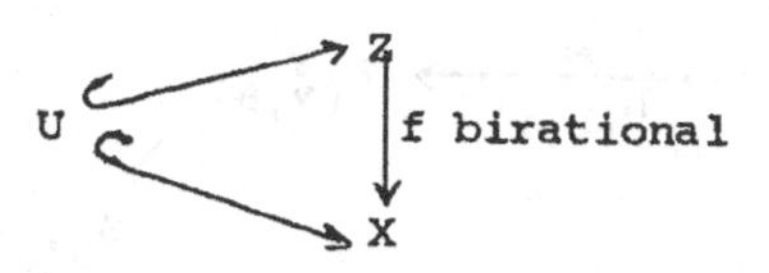

where:

1. Z has an open covering $\{V_i\}$ such that $U \subset V_i$, $f(V_i) \subset \text{Star } Y_i$ for some stratum Y_i and V_i is affine and canonical over Star Y_i

2. Z normal

We call these allowable modifications of X.

As far as I know, every normal canonical modification may be allowable. In any case we can construct them from f.r.p.p. decompositions Δ' of Δ by reversing the procedure followed at the end of §1: set

$$Z_{\Delta'} = \coprod_{\beta} Z_{\sigma'_\beta} \Big/ \left\{\begin{array}{l}\text{modulo equivalence relation} \\ \text{generated by open immersions} \\ Z_{\sigma'_\beta} \hookrightarrow Z_{\sigma'_\gamma} \quad \text{whenever } \sigma'_\beta = \text{face of } \sigma'_\beta\end{array}\right\}.$$

Explicitly, the equivalence relation is:

$$x_1 \in Z_{\sigma'_\beta} \text{ eq. to } x_2 \in Z_{\sigma'_\gamma} \iff \left\{\begin{array}{l}\text{the strata containing } x_1 \text{ and } x_2 \\ \text{correspond to a common face } \tau \text{ of} \\ \sigma'_\beta \text{ and } \sigma'_\gamma \text{ and } x_1 \text{ and } x_2 \text{ come from} \\ \text{the same point of } Z_\tau\end{array}\right\}$$

so that $Z_{\Delta'} = \left\{\text{disj. union of the closed strata } Y_\beta \text{ in each } Z_{\sigma'_\beta}\right\}$. The identification can be carried out like this: $\forall\ \beta,\gamma \quad \sigma'_\beta \cap \sigma'_\gamma$ is a closed subpolyhedron of σ'_β; hence it corresponds to a set of strata forming an open subscheme $Z_{(\beta,\gamma)} \subset Z_{\sigma'_\beta}$. Define

$$\begin{array}{ccc} Z_{(\beta,\gamma)} & \xrightarrow[h_{(\beta,\gamma)}]{\approx} & Z_{(\gamma,\beta)} \\ \cap & & \cap \\ Z_{\sigma'_\beta} & & Z_{\sigma'_\gamma} \end{array}$$

by requiring that for all faces $\tau \subset \sigma'_\beta \cap \sigma'_\gamma$, $h_{(\beta,\gamma)}$ should be given on the image of Z_τ by:

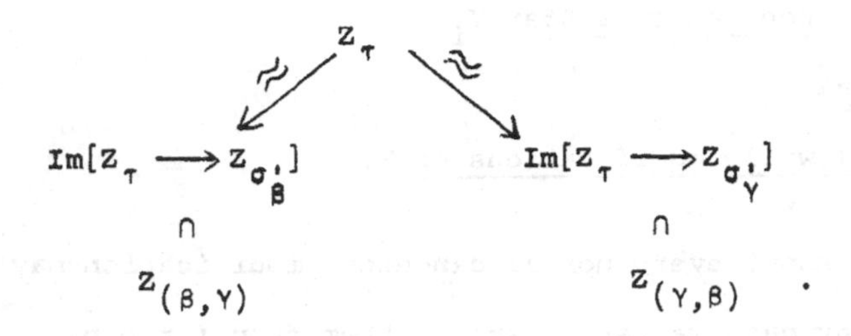

The only compatibility condition here is that when $\tau_1 \subseteq \tau_2 \subseteq \tau_3$, then the middle triangle here:

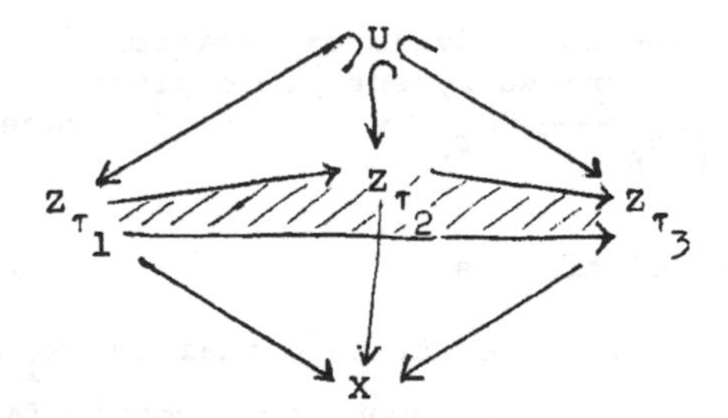

should commute. This is clear.

It is immediate from Th. 1* and this construction that (1) $Z_{\Delta'}$ fits into a diagram:

that (2) $U \hookrightarrow Z_{\Delta'}$ is a toroidal embedding without self-intersection, that (3) Δ' is the polyhedral complex associated to $U \hookrightarrow Z_{\Delta'}$, that (4) the diagram:

$$\begin{array}{ccc} R.S.^U(Z_{\Delta'}) & \xrightarrow{\text{ord}} & |\Delta'| \\ \cap & & \downarrow \\ R.S.^U(X) & \xrightarrow{\text{ord}} & |\Delta| \end{array}$$

commutes and that (5) if $\lambda \in R.S.^U(X)$, then $\lambda \in R.S.^U(Z_{\Delta'})$ iff ord $\lambda \in |\Delta'|$. By the valuative criterion for separation, $Z_{\Delta'}$ is separated (i.e., each $Z_{\sigma'_\beta}$ is affine over the separated scheme X, hence is separated; and if λ: Spec $k((t)) \longrightarrow U$ has extensions to 2 of these open pieces:

$$\mu: \text{ Spec } k[[t]] \longrightarrow Z_{\sigma'_\beta}$$

$$\nu: \text{ Spec } k[[t]] \longrightarrow Z_{\sigma'_\gamma}$$

then in $|\Delta|$, ord μ = ord ν, so $\mu(o), \nu(o)$ both lie in some Z_τ, τ = common face of $\sigma'_\beta, \sigma'_\gamma$ and μ, ν both factor through Z_τ.) This proves:

Theorem 6*: The correspondence $\Delta' \longmapsto Z_{\Delta'}$ defines a bijection between the f.r.p.p. decompositions of Δ and the isomorphism classes of allowable modifications of X.

We also find:

Theorem 7*: Let Δ', Δ'' be 2 f.r.p.p. decompositions of Δ. Then there exists a morphism g:

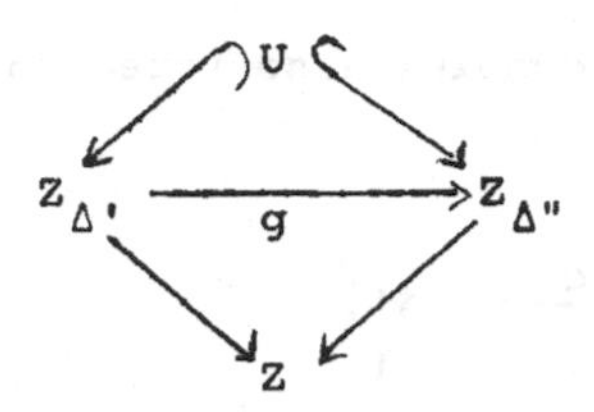

if and only if for all polyhedra σ'_β of Δ', $\sigma'_\beta \subseteq \sigma''_\gamma$ for some polyhedron σ''_γ of Δ''.

Theorem 8*: With the notation of Th. 7*, g is proper if and only if $|\Delta'| = |\Delta''|$.

(Proof immediate by valuative criterion for properness.)

Next we can generalize Th. 9 along the lines already indicated in Lemma 3. We consider canonical coherent sheaves $\mathfrak{J}$ of fractional ideals on X, i.e., $\forall \alpha$: $\hat{\mathfrak{O}}_{x_1,X} \xrightarrow{\approx} \hat{\mathfrak{O}}_{x_2,X}$ preserving strata, $\alpha(\hat{\mathfrak{J}}_{x_1}) = \hat{\mathfrak{J}}_{x_2}$. By Lemma 3, it follows that for any stratum Y,

$$\mathfrak{J}|_{\text{Star } Y} \cong \sum_{i=1}^{n} \mathfrak{O}_{\text{Star } Y}(-D_i), \quad \text{some } D_1, \cdots, D_n \in M^Y.$$

This allows us to define a map

$$\text{ord}\,\mathfrak{J}: \; |\Delta| \longrightarrow \mathbb{R}$$

in the following way:

$$\forall\, x \in \sigma^Y, \qquad \text{ord}\,\mathfrak{J}(x) = \min_{1\le i\le n} \langle D_i, x\rangle .$$

Note that:

$$\begin{aligned}\text{ord}\,\mathfrak{J}(\text{ord}\,\lambda) &= \min_{1\le i\le n} \langle D_i, \text{ord}\,\lambda\rangle \\ &= \min_{1\le i\le n} \text{ord}_o \lambda^{-1}(D_i) \\ &= \text{ord}_o \lambda^{-1}(\mathfrak{J})\end{aligned}$$

and hence the definition is independent of the choice of D_i. Clearly ord $\mathfrak{J}$ is a function $f: |\Delta| \longrightarrow \mathbb{R}$ such that:

(*)

(i) $f(\lambda\cdot x) = \lambda\cdot f(x)$, $\lambda \in \mathbb{R}^+$,

(ii) f is continuous, piecewise-linear,

(iii) $f(\sigma^Y \cap N^Y) \subset \mathbb{Z}$, all Y,

(iv) f is convex on each σ^Y.

Conversely let $f: |\Delta| \longrightarrow \mathbb{R}$ satisfy conditions (*). For all Y, put

$$(\mathfrak{J}_f)_Y = \sum_{\substack{D\in M^Y \\ D\ge f \text{ on } \sigma^Y}} \mathcal{O}_{\text{Star }Y}(-D) .$$

Theorem 9*:

I. Let $f: |\Delta| \longrightarrow \mathbb{R}$ satisfy (*). Then the $(\mathfrak{I}_f)_Y$ can be patched together into a canonical coherent complete sheaf $\mathfrak{I}_f$ of fractional ideals on X.

II. a) $\operatorname{ord} \mathfrak{I}_f = f$

b) $\mathfrak{I}_{\operatorname{ord} f}$ is the completion of f

c) The maps $\mathfrak{I} \longmapsto \operatorname{ord} \mathfrak{I}$ and $f \longmapsto \mathfrak{I}_f$ define a bijection between the set of canonical coherent complete sheaves of fractional ideals and the set of functions f satisfying (*).

d) $\mathfrak{I} \subset \mathfrak{I}_f$ iff $\operatorname{ord} \mathfrak{I} \geq f$,

e) $\operatorname{ord} \mathfrak{I}_1 \cdot \mathfrak{I}_2 = \operatorname{ord} \mathfrak{I}_1 + \operatorname{ord} \mathfrak{I}_2$

f) $\mathfrak{I}|_{\operatorname{Star} Y} \cong \mathcal{O}_{\operatorname{Star} Y}$ iff $\operatorname{ord} \mathfrak{I} \equiv 0$ on σ^Y

III. (Notation as in Th. 8, Ch. I)

a) $\mathfrak{I}^{-1} = \mathfrak{I}_g$ where g is the convex interpolation of $-\operatorname{ord} \mathfrak{I}$ on $\bigcup_Y \operatorname{Sk}^1(\sigma^Y)$.

b) $(\mathfrak{I}^{-1})^{-1} = \mathfrak{I}$ if and only if $\mathfrak{I}$ is complete and $\operatorname{ord} \mathfrak{I}$ is the convex interpolation of a function $\bigcup_Y \operatorname{Sk}^1(\sigma^Y) \longrightarrow \mathbb{Z}$.

Moreover there exists a bijective correspondence between the set of canonical Weil-divisors (i.e., those supported on X-U) and the set of integral functions on $\bigcup_Y \operatorname{Sk}^1(\sigma^Y)$.

c) The following are equivalent:

i) $\mathfrak{I}$ invertible

ii) $\mathfrak{I} \cdot \mathfrak{I}^{-1} = \mathcal{O}_X$

iii) $\operatorname{ord} \mathfrak{I}$ linear on each σ^Y

The Proof is similar to that of Th. 9, Ch. I.

Theorem 10*: Let $\mathfrak{J}$ be a canonical coherent sheaf of fractional ideals. Let $B_{\mathfrak{J}}(X)$ be the normalization of the variety obtained by blowing up $\mathfrak{J}$. Then $B_{\mathfrak{J}}(X)$ is an allowable modification of X and is described by the f.r.p.p. decomposition of Δ obtained by subdividing the σ^Y's into the biggest possible polyhedra on which ord $\mathfrak{J}$ is linear.

Proof: First, $f: B_{\mathfrak{J}}(X) \longrightarrow X$ is an allowable modification: in fact, if

$$\mathfrak{J}|_{\text{Star } Y} \cong \sum_{i=1}^{n} \mathfrak{O}_{\text{Star } Y}(-D_i),$$

then $f^{-1}(\text{Star } Y)$ is covered by the n relatively affine open pieces which are the normalizations of:

$$V_i = \underset{\sim}{\text{Spec}}\, \mathcal{A}_i, \quad \mathcal{A}_i = \left\{ \mathfrak{O}_{\text{Star } Y}\text{-algebra generated by } \mathfrak{O}(D_i - D_j),\ 1 \leq j \leq n \right\}$$

[since if, locally in Spec $R \subset$ Star Y, δ_i is an equation of D_i, then $\mathfrak{J}$ is given by the fractional ideal $\sum \delta_i R$, hence the blow-up is covered by affines with rings

$$S_i = R\left[\frac{\delta_1}{\delta_i}, \ldots, \frac{\delta_n}{\delta_i}\right]$$

and S_i is the $\mathfrak{O}_{\text{Star } Y}$-algebra generated by $\mathfrak{O}(D_i - D_1), \cdots, \mathfrak{O}(D_i - D_n)$.]
As $B_{\mathfrak{J}}(X)$ is also characterized as the minimal normal variety dominating X such that the pull-back of $\mathfrak{J}$ is invertible, Th. 10* follows from Th. 7* and Th. 9* (Part IIIc). QED

Finally the proof of Th. 11 goes over immediately to prove:

<u>Theorem 11*</u>: <u>For any toroidal embedding</u> $U \subset X$ <u>without self-intersection</u>, <u>there exists a canonical sheaf of ideals</u> $\mathcal{I} \subset \mathcal{O}_X$ <u>such that</u> $B_{\mathcal{I}}(X)$ <u>is non-singular</u>.

There is another situation that we must analyze for the sake of its application to semi-stable reduction. This does not involve any new ideas but rather a slight reformulation of what has been studied so far in a new situation. Suppose that in addition to $U \subset X$, a toroidal embedding without self-intersection, we are given a positive Cartier-divisor D with support exactly X-U. We can associate to the triple (X,U,D) a <u>compact</u> polyhedral complex Δ_o with integral structure, where

$$|\Delta_o| = \left\{ x \in |\Delta| \,\middle|\, \langle D,x\rangle = 1 \right\},$$

$$\sigma_o^Y = |\Delta_o| \cap \sigma^Y,$$

$$\mathrm{res}_{\sigma_o^Y}(M^Y) \text{ giving integral structure.}$$

Note that a polyhedral subdivision of Δ_o gives a <u>conical</u> polyhedral subdivision of Δ and vice versa. When one has a compact polyhedral complex with integral structure $(\Delta_o, \sigma_\alpha, L_\alpha)$, note that one can define several more structures on Δ_o:

a) an increasing series of "lattices" on Δ_o:

$\nu \geq 1$, let

$$(\Delta_o)_{\frac{1}{\nu}\mathbb{Z}} = \left\{ x \in |\Delta_o| \,\middle|\, \begin{matrix} \text{if } x \in \sigma_\alpha, \text{ then for all} \\ f \in L_\alpha,\ f(x) \in \frac{1}{\nu}\mathbb{Z} \end{matrix} \right\}$$

Every rational point of Δ_o lies on one of these lattices but each $(\Delta_o)_{\frac{1}{\nu}\mathbb{Z}}$ is a finite set.

b) a volume element on each polyhedron σ_α or even on each rational polyhedron $\tau \subseteq \sigma_\alpha$ (possibly of lower dimension than σ_α): let $L = \mathrm{res}_\tau L_\alpha$ and if $k = \dim \tau$, let $\frac{1}{a}, f_1, \cdots, f_k$ be a basis of L. Use $f_1, \cdots, f_k$ to define $F: \tau \longrightarrow \mathbb{R}^k$ and pull-back the volume element. Since this embedding is unique up to translations and unimodular transformations, the volume element is well-defined.

Theorem 12*: Given $U \subset X$ and D as above, let Δ' be an f.r.p.p. decomposition of Δ and let Δ'_o be the associated decomposition of Δ_o. Let $f: Z_{\Delta'} \longrightarrow X$ be the corresponding modification. Then

a) the vertices of Δ'_o are in $(\Delta_o)_{\mathbb{Z}}$ iff $f^{-1}(D)$ vanishes to order one on each component of $Z_{\Delta'} - U$,

b) If (a) holds, then moreover the volume of every polyhedron τ_o in Δ'_o is $1/(\dim \tau_o)!$ iff $Z_{\Delta'}$ is non-singular.

Proof: To prove (a), note that the components E_i of $Z_{\Delta'} - U$ correspond to the one-dimensional faces $\mathbb{R}^+ \cdot v_i$ of Δ' and hence to the vertices of Δ'_o. If we normalize v_i so that v_i is a primitive vector in N^Y, then $\langle D, v_i\rangle \in \mathbb{Z}$. Let $\langle D, v_i\rangle = \nu$. Then $\frac{1}{\nu} v_i$ is the corresponding vertex of Δ'_o and

$$\frac{1}{\nu}v_i \in (\Delta_o)_{\mathbb{Z}} \quad \text{iff} \quad \nu = 1.$$

On the other hand

$$\begin{aligned}\nu &= \text{least integer } n \text{ such that } \langle D,v\rangle = n, \text{ some} \\ &\qquad v \in N^Y \cap (\mathbb{R}^+ \cdot v_i) \\ &= \text{least integer } n \text{ such that } \langle D, \text{ord } \lambda\rangle = n, \text{ some} \\ &\qquad \lambda \in R.S.^U(X) \text{ with } \lambda(0) \in E_i - \bigcup_{j\neq i} E_j \\ &= \text{multiplicity to which } f^{-1}(D) \text{ vanishes along } E_i.\end{aligned}$$

This proves (a). As for (b) note that if τ is a k-dimensional compact polyhedron with integral vertices, then

$$\text{vol}(\tau) = \frac{a}{k!}, \quad a \in \mathbb{Z}, \quad a \geq 1$$

and that certainly $a > 1$ unless τ is a simplex. Thus for either of the 2 conditions in (b) to hold, all τ_o must be simplices. We are now reduced to:

<u>Lemma 4</u>: <u>Let</u> $N_{\mathbb{R}}$ <u>be a real vector space,</u> $N \subset N_{\mathbb{R}}$ <u>a lattice,</u> $N^* = \text{Hom}(N,\mathbb{Z})$. <u>Let</u> $x_o,\cdots,x_k \in N$ <u>be independent vectors such that</u> $\exists \ell \in N^*$ <u>with</u> $\ell(x_i) = 1$, $D \leq i \leq k$ <u>and let</u> $\tau = [\text{convex hull of } x_o,\cdots,x_k$ <u>in the hyperplane</u> $\ell = 1]$. <u>Using</u> N^*, <u>induce a volume element on</u> τ. <u>Then</u> (<u>in the notation of</u> Ch. I, §2):

$$\text{vol}(\tau) = \frac{\text{mult}\langle x_o,\cdots,x_k\rangle}{k!}$$

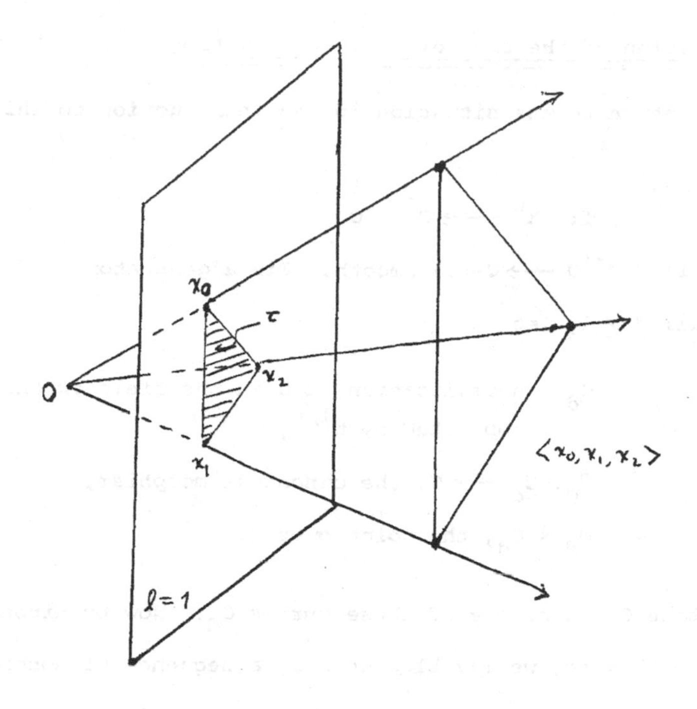

<u>Proof</u>: Choose an isomorphism $N \cong \mathbb{Z}^{k+1}$ so that $x_o = (1,0,\cdots,0)$ and $\ell(a_o,\cdots,a_k) = a_o$. If $x_i = (1,a_1^{(i)},\cdots,a_k^{(i)})$, then

$$\text{mult}\langle x_o,\cdots,x_k\rangle = \det\begin{pmatrix} 1 & 0 & \cdots & 0 \\ 1 & a_1^{(1)} & \cdots & a_k^{(1)} \\ 1 & a_1^{(k)} & \cdots & a_k^{(k)} \end{pmatrix}$$

and

$$\text{vol}(\tau) = \frac{1}{k!}\det\begin{pmatrix} a_1^{(1)} & \cdots & a_k^{(1)} \\ a_1^{(k)} & \cdots & a_k^{(k)} \end{pmatrix} .$$

<u>QED</u>

§3. Reduction of the theorem to a construction

Now return to the situation in the introduction to this chapter:

$$f: X^n \longrightarrow C^1 \ni 0$$

with res $f: X-f^{-1}0 \longrightarrow C-(0)$ smooth. Fix a generator t of $m_{0,C}$, and for all $d \geq 1$, let

C_d = normalization of C in the field extension generated by $t^{1/d}$,

$\pi_d: C_d \longrightarrow C$, the canonical morphism,

$0_d \in C_d$, the point over 0.

We will take C' to be one of these curves C_d. Now by Hironaka's resolution theorem, we may blow up X by a sequence of monoidal transformations with non-singular centers all lying over $0 \in C$, until we find:

$$g: X' \longrightarrow X \quad \text{birational and projective}$$

with

X' non-singular,

$(f\cdot g)^{-1}(0)_{red}$ a union of non-singular components $E_1,\cdots,E_N$ crossing transversely.

The only problem is that if $n(i) = \mathrm{ord}_{E_i}(t)$ so that

$$(f\cdot g)^{-1}(0) \underset{\text{as divisor on } X'}{=\!=} \sum n(i)E_i,$$

then the $n(i)$ may be bigger than one, hence $(f\cdot g)^{-1}(0)$ may not be reduced. To reduce the $n(i)$, we must replace c by a suitable C_d. It is clear that we may as well rename X' to be the original X and assume that our starting point is a non-singular X with $f^{-1}(0)_{red}$ good. Now for all $d \geq 1$, let

$$X_d = \text{normalization of } X \times_C C_d$$

$$f_d : X_d \longrightarrow C_d \quad \text{the projection}$$

$$U_d = f_d^{-1}(C_d - (0_d)).$$

What does X_d look like? To describe it, choose a point $x \in X$ over $0 \in C$. Suppose $x \in E_{i_1} \cap \cdots \cap E_{i_r}$ and $x \notin$ all other E_j. Then formally at x,

$$\left\{\begin{matrix} X & x \\ \downarrow & \downarrow \\ C & 0 \end{matrix}\right\} \text{ isomorphic to } \left\{\begin{matrix} 0 & \in \mathbb{A}^n \\ \downarrow & \downarrow P \\ 0 & \in \mathbb{A}^1 \end{matrix}, \quad \text{given by } t = \prod_{j=1}^{r} x_j^{n(i_j)}\right\}$$

This is where we use characteristic 0! In fact, if $y_j \in \mathcal{O}_{x,X}$ is a local equation for the divisor E_{i_j}, then

$$t = u \cdot \prod y_j^{n(i_j)}, \qquad u \in \mathcal{O}_{x,X}^*.$$

In the completion $\hat{\mathcal{O}}_{x,X}$, $y_1, \cdots, y_r$ are part of a system of parameters since the E_{i_j} meet transversely. Moreover, since char $k \nmid n(i_1)$, u has an $n(i_1)^{th}$ root v in $\hat{\mathcal{O}}_{x,X}^*$. So we may replace y_1 by $v \cdot y_1$, and find local equations $y_j' \in \hat{\mathcal{O}}_{x,X}$ of E_{i_j} such that $t = \prod y_j'^{n(i_j)}$, i.e.,

$$\hat{\mathcal{O}}_{x,X} \cong k[[y_1', \cdots, y_n']]$$

$$t \longleftrightarrow \prod_{j=1}^{r} y_j'^{n(i_j)}$$

Therefore, formally at all points over x:

$$\left\{\begin{array}{ccc} X_d & \longrightarrow & X \\ \downarrow & & \downarrow \\ C_d & \longrightarrow & C \end{array}\right\} \text{ isomorphic to } \left\{\begin{array}{ccc} \text{norm. of } \mathbb{A}^n \times_{\mathbb{A}^1} \times \mathbb{A}^1 & \longrightarrow & \mathbb{A}^n \\ \downarrow & & \downarrow p \\ \mathbb{A}^1 & \xrightarrow[d^{th} \text{ power}]{} & \mathbb{A}^1 \end{array}\right\}$$

(Note that normalization commutes with taking completions.)

Let $s = t^{1/d}$. Then

$$\mathbb{A}^n \times_{\mathbb{A}^1} \mathbb{A}^1 = \left[\text{the hypersurface H: } s^d = \prod x_j^{n(i_j)} \text{ in } \mathbb{A}^{n+1}\right].$$

Let

$$e = \text{g.c.d.}(d, n(i_1), \cdots, n(i_r)).$$

Then

$$H = \bigcup_{\substack{e^{th} \text{ roots} \\ \text{of } 1}} H_\zeta$$

$$H_\zeta = \left[\text{hypersurface } s^{d/e} = \zeta \cdot \prod_{j=1}^{r} x_j^{n(i_j)/e}\right].$$

H_ζ is the image of the morphism

$$\mathbb{A}^n_{\tilde{x}} \longrightarrow \mathbb{A}^{n+1}_{(x,s)}$$

$$x_i = (\tilde{x}_i)^d$$

$$s = \zeta^{e/d} \cdot \prod_{j=1}^{r} \tilde{x}_j^{n(i_j)}$$

hence is irreducible, and its coordinate ring is the subring:

$$k[x_1,\cdots,x_n,\prod_{j=1}^{r} x_j^{n(i_j)/d}] \subset k[x_1^{1/d},\cdots,x_n^{1/d}].$$

Thus if we let

$$M = \mathbb{Z}^n + (\frac{n(i_1)}{d},\cdots,\frac{n(i_r)}{d},0,\cdots,0)\mathbb{Z} \subset \mathbb{Q}^n,$$

then H_ζ is isomorphic to the affine embedding of the torus T with character group M, given by the semi-group in M generated by:

$$\begin{gathered}(1,0,\cdots,0)\\ \cdots\cdots\\ (0,0,\cdots,1)\\ (\frac{n(i_1)}{d},\cdots,\frac{n(i_r)}{d},\cdots,0).\end{gathered}$$

The cone generated by this semi-group in $M_{\mathbb{R}} \cong \mathbb{R}^n$ is just the positive octant $(\mathbb{R}_+)^n$, so we can normalize as in Ch. I, §1:

$$\begin{bmatrix}\text{normalization}\\ \text{of } H_\zeta\end{bmatrix} \cong \operatorname{Spec} k[\cdots\cdots,x^\alpha,\cdots\cdots]_{\alpha\in M\cap(\mathbb{R}^+)^n}.$$

Going back, this means that $x \in X$ splits into e distinct points $x' \in X_d$ and that at each of them:

$$\hat{\mathcal{O}}_{x',X_d} \cong k[[\cdots,y^\alpha,\cdots]]_{\alpha\in M\cap(\mathbb{R}_+)^n}.$$

Under this formal isomorphism:

$$\begin{aligned}\sum E_i \quad &\text{given by } \prod y_i = 0\\ &\text{corresp. to } \prod x_i = 0\\ &\text{defining } H_\zeta - T.\end{aligned}$$

Therefore $U_d \subset X_d$ is a toroidal embedding! Moreover the components of $X_d - U_d$ are the components of the inverse image in X_d of the various $E_i \subset X$. The inverse image of E_{i_j} near x', with reduced structure, is given by $\mathcal{O}_{x',X_d}/\sqrt{(y_j)}$, and

$$\widehat{\mathcal{O}_{x',X_d}/\sqrt{(y_j)}} \cong k[[\cdots,y^\alpha,\cdots]]_{\alpha\in M\cap\mathbb{R}^n_+}\Big/\sqrt{(y_j)}$$

$$\cong k[[\cdots,y^\alpha,\cdots]]_{\alpha\in M\cap\mathbb{R}^n_+}\Big/(\cdots,y^\alpha,\cdots)_{\alpha_j>0}$$

$$\cong k[[\cdots,y^\alpha,\cdots]]_{\alpha_j=0,\alpha\in M\cap\mathbb{R}^n_+} .$$

This again is an integrally closed domain of toroidal type. Thus the inverse image of each E_i is a disjoint union of normal varieties. This proves:

Lemma 1: $U_d \subset X_d$ is a toroidal embedding without self-intersection.

The next question is: how does X_d vary with d? Let $\nu = \text{l.c.m.}\,(n(1),\cdots,n(N))$. We are only interested in d such that $\nu | d$. Suppose $d = e.\nu$ and consider $\hat{\mathcal{O}}_{x',X_d}$ again at a random point x. Then

$$n(i_j)\cdot m(i_j) = \nu, \quad 1 \leq j \leq r$$

and

$$M = \mathbb{Z}^n + \left(\frac{1}{e\cdot m(i_1)},\cdots,\frac{1}{e\cdot m(i_r)},0,\cdots,0\right)\mathbb{Z} .$$

Let

$$M_o = \mathbb{Z}^n + \left(\frac{1}{m(i_1)},\cdots,\frac{1}{m(i_r)},0,\cdots,0\right)\mathbb{Z} .$$

Suppose $\alpha \in M \cap (\mathbb{R}_+)^n$. Then either:

1) $\alpha_1,\cdots,\alpha_r > 0$, in which case

$$\alpha = (\frac{1}{e\cdot m(i_1)},\cdots,\frac{1}{e\cdot m(i_r)},0,\cdots,0) + \beta, \quad \beta \in M \cap (\mathbb{R}_+)^n,$$

or

2) some $\alpha_\ell = 0$, $1 \leq \ell \leq r$. Now

$$\alpha = \text{integral vector} + p\cdot(\frac{1}{e.m(i_1)},\cdots,\frac{1}{e\cdot m(i_r)},0,\cdots,0)$$

so $e\cdot m(i_\ell)\,|\,p$. Thus $e|p$, and

$$\alpha \in M_o \cap \mathbb{R}^n_+.$$

This means that the semi-group $M \cap \mathbb{R}^n_+$ is generated by $M_o \cap \mathbb{R}^n_+$ and the vector $(\frac{1}{e\cdot m(i_1)},\cdots,\frac{1}{e\cdot m(i_r)},0,\cdots,0)$. In terms of rings, this means that $\hat{\mathcal{O}}_{x',X_d}$ is generated by $\hat{\mathcal{O}}_{x',X_\nu}$ and $t^{1/d}$. Therefore the canonical morphism

$$X_d \longrightarrow X_\nu \times_{C_\nu} C_d$$

induces isomorphisms between the complete local rings of corresponding points, hence it is étale. But it is also finite and birational, hence it is an isomorphism. This proves:

<u>Lemma 2</u>*: <u>If $\nu|d$, then $X_d \xrightarrow{\approx} X_\nu \times_{C_\nu} C_d$. Hence the closed fibres $f_d^{-1}(o_d)$ is independent of d so long as $\nu|d$, and the projection</u>

---

*In fact, one also checks easily that $t^{1/d}$ vanishes to order 1 on all components of $X_d - U_d$, hence $f_d^{-1}(o_d)$ is also a reduced scheme. But we don't actually use this particular fact.

$X_\nu \longrightarrow X_d$ <u>induces a bijection between the strata of</u> $X_\nu - U_\nu$ <u>and the strata of</u> $X_d - U_d$.

Next, let Δ_d be the polyhedral complex associated to $U_d \subset X_d$. I claim that there is a canonical polyhedral isomorphism $\Delta_d \xrightarrow{\approx} \Delta_\nu$ and in fact a commutative diagram:

$$\begin{array}{ccc} R.S.^{U_d}(X_d) & \xrightarrow{\text{ord}} & \Delta_d \\ \downarrow & & \downarrow \wr \\ R.S.^{U_\nu}(X_\nu) & \xrightarrow{\text{ord}} & \Delta_\nu \end{array}$$

(the first vertical arrow given by composing ϕ: Spec $k[[t]] \longrightarrow X_d$ with the projection $p: X_d \longrightarrow X_\nu$). In fact, let $Y_d \subset X_d$ and $Y_\nu \subset X_\nu$ be corresponding strata as in Lemma 2. Then $p^{-1}(\text{Star } Y_\nu) = \text{Star } Y_d$ and we get homomorphisms:

$$M^{Y_d} \underset{p_*}{\overset{p^*}{\rightleftarrows}} M^{Y_\nu}$$

where p^* = pull-back of Cartier divisors, p_* = Norm (i.e., apply norm to local defining equations). Then $p_* \cdot p^*$ = mult. by e. Moreover p_* is injective (in fact, since p is a bijection on $f_d^{-1}(o_d)$, p_* is injective on Weil divisors concentrated on $f_d^{-1}(o_d)$). Thus $p^* \cdot p_*$ = mult. by e too and

$$M_{\mathbb{R}}^{Y_d} \xleftarrow[p^*]{\approx} M_{\mathbb{R}}^{Y_\nu}$$

is an isomorphism. Clearly p^*D is effective if and only if D is effective, so this induces a dual isomorphism

$$\sigma^{Y_d} \xrightarrow{\approx} \sigma^{Y_\nu}.$$

These clearly patch up into an isomorphism $\Delta_d \xrightarrow{\approx} \Delta_\nu$ commuting with ord.

However, the one thing which changes when you replace Δ_ν by Δ_d is the integral structure. The integral structures on corresponding polyhedra $\sigma^{Y_d}, \sigma^{Y_\nu}$ are given by the functions defined by M^{Y_d} and M^{Y_ν} respectively. I claim:

$$M^{Y_d} = p^*M^{Y_\nu} + \mathbb{Z}\cdot(t^{1/d})$$

or equivalently:

<u>Lemma 3</u>: <u>Every Cartier divisor</u> D <u>on Star</u> Y_d, <u>supported by</u> $f_d^{-1}(o_d)$, <u>is of the form</u> $p^*D_1 + a\cdot(t^{1/d})$, $a \in \mathbb{Z}$.

<u>Proof</u>: In the notation used above, the morphism $p\colon X_d \longrightarrow X_\nu$ corresponds formally, at every $x' \in X_d$, to the morphism of affine torus embeddings:

$$\operatorname{Spec} k[\cdots, x^\alpha, \cdots]_{\alpha \in M \cap \mathbb{R}^n_+} \longrightarrow \operatorname{Spec} k[\cdots, x^\alpha, \cdots]_{\alpha \in M_o \cap \mathbb{R}^n_+}$$

But if we choose $x' \in Y_d$, then the formal isomorphisms induce isomorphisms $M^{Y_d} \cong M$, $M^{Y_\nu} \cong M_o$ which lie in a diagram:

$$\begin{array}{ccc} M^{Y_d} & \cong & M \\ \uparrow p^* & & \cup \\ M^{Y_\nu} & \cong & M_o \, . \end{array}$$

Also, (t) corresponds in these isomorphisms to $\alpha_o = (n(i_1),\cdots,n(i_r),0,\cdots,0)$. Since $M = M_o + \frac{\alpha_o}{d}\cdot\mathbb{Z}$, this proves Lemma 3.

Finally, in all the toroidal embeddings $U_d \subset X_d$, we are given a particular positive Cartier divisor with support exactly $X_d - U_d$, namely $(t^{1/d})$. Let $(t^{1/d})$ define the function $\ell_d\colon \Delta_d \longrightarrow \mathbb{R}_+$. Note that via $\Delta_d \xrightarrow{\approx} \Delta_\nu$, $\ell_d = \frac{\nu}{d}\ell_\nu$. As in the end of §2, we can therefore define a compact polyhedral complex

$$\Delta_d^* = \left\{ x \in \Delta_d \,\middle|\, \ell_d(x) = 1 \right\}.$$

By restriction, we get an integral structure M_d^* on Δ_d^*. Moreover, by central projection and the canonical isomorphism $\Delta_d \xrightarrow{\approx} \Delta_\nu$, we get a canonical isomorphism

$$\Delta_d^* \xrightarrow{\approx} \Delta_\nu^*,$$

and by Lemma 3, the integral structures are related by

$$M_d^* = \frac{d}{\nu}M_\nu^* + \mathbb{Z}\,.$$

In other words, via these isomorphisms:

$$\left[\begin{array}{l}\text{the integral lattice}\\ (\Delta_d^*)_{\mathbb{Z}} \text{ in } \Delta_d^*\end{array}\right] = \left[\begin{array}{l}\text{the lattice of } (\Delta_\nu^*)_{\frac{\nu}{d}\mathbb{Z}}\\ \text{of points in } \Delta_\nu^* \text{ with}\\ \text{coordinates in } \frac{\nu}{d}\mathbb{Z}\end{array}\right]$$

Now the Main Theorem of the next Chapter, applied to Δ_ν^*, says that there is an integer e and a projective subdivision $\{\sigma_\alpha\}$ of Δ_ν^* such that:

a) vertices of the subdivision lie in $(\Delta_\nu^*)_{\frac{1}{e}\mathbb{Z}}$,

b) for all σ_α, with the volume element induced from Δ_ν^*,

$$\mathrm{vol}(\sigma_\alpha) = \frac{1}{e^{d_\alpha}\cdot(d_\alpha)!}, \quad d_\alpha = \dim \sigma_\alpha .$$

It follows that if we interpret this as a subdivision of $\Delta_{e\nu}^*$ instead of Δ_ν^*, then:

a') vertices of the subdivision lie in $(\Delta_{e\nu}^*)_{\mathbb{Z}}$

b') for all σ_α of dim = d_α, $\mathrm{vol}(\sigma_\alpha) = 1/(d_\alpha)!$

Now apply the results of §2: $\{\sigma_\alpha\}$ defines a proper birational morphism:

$$f: X_{\{\sigma_\alpha\}} \longrightarrow X_{e\nu}, \quad f^{-1}(U_{e\nu}) \cong U_{e\nu}.$$

Since the subdivision is projective, the morphism is defined by blowing up a suitable sheaf of ideals. By Theorem 1*, $t^{1/e\nu}$ vanishes to order 1 on all components of $X_{\{\sigma_\alpha\}}-U_{e\nu}$, and $X_{\{\sigma_\alpha\}}$ is non-singular. It follows from this automatically that the components of $X_{\{\sigma_\alpha\}}-U_{e\nu}$ are non-singular and cross transversely (because $U_{e\nu} \subset X_{\{\sigma_\alpha\}}$ is a toroidal embedding without self-intersection). Therefore $X_{\{\sigma_\alpha\}}$ has all the required properties.

CHAPTER III

Construction of nice polyhedral subdivisions

Finn F. Knudsen

I don't know whether from a combinatorial point of view the following question has ever been asked:

Given a polyhedron $\sigma \subset \mathbb{R}^n$ with integral vertices, find an integer $\nu \geq 1$ and a decomposition of σ into simplices τ_α such that for all α:

1) vertices of $\tau_\alpha \subset \frac{1}{\nu}\mathbb{Z}^n$

2) volume $(\tau_\alpha) = \dfrac{1}{\nu^n n!}$.

But the theory of the previous chapter makes it clear that this is the essential construction needed to carry out semi-stable reduction. The purpose of this Chapter is to study and solve this combinatorial problem with certain refinements ("projectivity of the subdivision in the case when σ is convex and globalization") independent of algebraic geometry.

One of the key steps (4.2) is due to Alan Waterman. The rest is a truly joint effort by Mumford and me.

§1. Definitions and Projective subdivisions

We continue to use the definition of a compact polyhedral complex X with an integral structure $M = \{M_\alpha\}$ (definitions 5 and 6 of §1 in the previous chapter) except that we call M a rational structure here. If μ is an integer ≥ 1 such that the functions in M take values in $\frac{1}{\mu}\mathbb{Z}$ on the vertices, then we say that M is integral over $\frac{1}{\mu}\mathbb{Z}$. For such μ, we have:

Definition 1.1. The number

$$m(\sigma_\alpha, M_\alpha, \mu) = \mu^{\dim(\sigma_\alpha)} \cdot (\dim \sigma_\alpha)! \cdot \mathrm{vol}\, \sigma_\alpha$$

is an integer, and we will call this integer the multiplicity of σ_α with respect to M and μ.

Also we define

$$m(X, M, \mu) = \max_{\sigma_\alpha \subset |X|} \{m(\sigma_\alpha, M_\alpha, \mu)\} .$$

Observation 1.2. Let $M \subset N$ be two rational structures on a polyhedral complex X, such that

$$M \cap \{\text{constants}\} = N \cap \{\text{constants}\} .$$

Let μ be an integer such that (X,M) and (X,N) are both integral over $\frac{1}{\mu}\mathbb{Z}$. Then for any polyhedron σ_α we have

$$m(\sigma_\alpha, M, \mu) = \# (N_\alpha / M_\alpha) \cdot m(\sigma_\alpha, N, \mu) .$$

Suppose M is integral over $\mathbb{Z}$ and contains the constant 1. Then

$$m(\sigma_\alpha, M, 1) = 1 \qquad \text{if and only if :}$$

σ_α is a simplex, and M_α is generated by the functions x_i, where x_i is the linear function on σ_α which is 1 on the vertex P_i and 0 on all the other vertexes.

Definition 1.3. Let X and X' be polyhedral complexes. We say that X' is a subdivision of X if

i) X and X' have the same underlying topological space,

ii) Whenever τ is a polyhedron of X', there is a polyhedron σ of X such that $\tau \subset \sigma$,

iii) If τ is in X' and σ in X and $\tau \subset \sigma$, then

$$V'_\tau = \text{res}_\tau V_\sigma .$$

Observation 1.4. For any σ in X let σ' be the topological space σ together with the collection of polyhedra τ in X' such that $\tau \subset \sigma$. Then it follows that σ' is a polyhedral complex and σ' is a subdivision of σ.

Definition 1.5. Let X' be a subdivision of a polyhedral complex X. We say that X' is a projective subdivision of X if there exists a real-valued continuous function f: X $\longrightarrow \mathbb{R}$ such that

i) For each polyhedron $\sigma \subset X$ $f|_\tau$ is piecewise-linear and convex, i.e., $f|_\tau = \min_{1 \leq i \leq N} \ell_i$, $\ell_i \in V_\sigma$.

ii) If σ is a polyhedron in X and ℓ a linear function on σ such that $\ell \geq f|_\sigma$, then the set

$$\tau = \left\{ x \in \sigma \mid f(x) = \ell(x) \right\}$$

is either empty or a polyhedron of X'.

Whenever X' is a subdivision of X and f is a continuous function which satisfies i) and ii) we will say that f is a good function for the subdivision X'.

Definition 1.6. Let X be a polyhedral complex with a rational structure M. We say that a subdivision X' of X is a rational subdivision if the functions in M take rational values on the vertexes of X'. When X' is a rational subdivision of X we restrict the rational structure to X'.

Definition 1.7. Let X be a polyhedral complex. Then we denote by PL(X) the set of real-valued continuous functions f on X such that $f|_\tau \in V_\tau$ for each polyhedron τ.

Note that evaluation of f at the vertexes of f gives us a linear embedding:

$$PL(X) \longrightarrow \mathbb{R}^{X_0}$$

Next we will prove a numerical criterion for projectiveness.

Lemma 1.8. *Let X' be a subdivision of a polyhedral complex X. Then we can find a finite collection of linear functions:*

$\Delta_i : PL(X')/PL(X) \longrightarrow \mathbb{R}$ *with the property that if f is a function in PL(X'), then f is a good function for the subdivision X' of X if and only if* $\Delta_i(t) > 0$ *for all i.*

Proof. Let σ be a polyhedron in X, say of dimension n, and let τ be an n-1 polyhedron of X' such that $int(\tau) \subset int(\sigma)$. There are exactly two n-polyhedra τ_1 and τ_2 of X' such that $\tau_i \subset \sigma$, $i = 1,2$ and $\tau = \tau_1 \cap \tau_2$. Let P,Q be points in int τ_1 and int τ_2 respectively such that the line segment P,Q meets τ in, say, a point R. For any $f \in PL(X')$ we define

$$\Delta_\tau(f) = f(R) - g(R)$$

where g is the linear function on the line segment P,Q such that $g(P) = f(P)$ and $g(Q) = f(Q)$

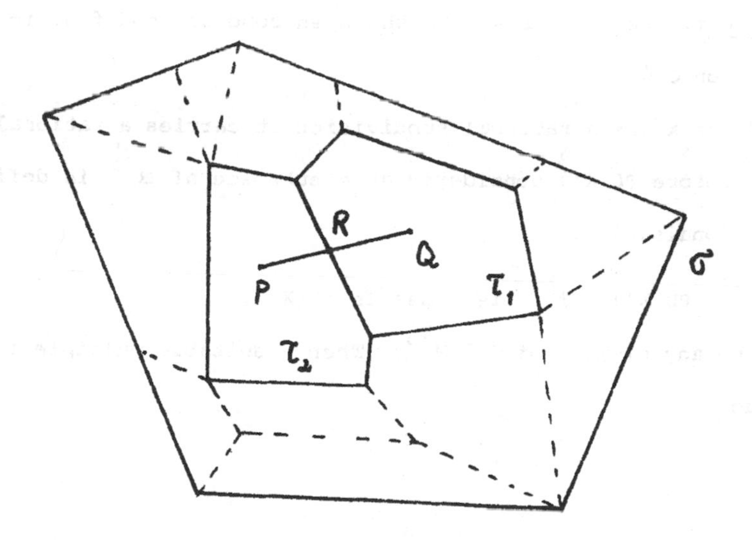

Now choose such a function Δ_τ for all τ with the above property, i.e., $\text{int}(\tau) \subset \text{int}(\sigma)$ where σ is a top dimensional polyhedron of X and $\dim \tau = \dim(\sigma)-1$.

Claim: $f \in PL(X')$ is good if and only if $\Delta_\tau(f) > 0$ for all such τ. We leave the proof of this to the reader.

We have two immediate corollaries.

Corollary 1.9. Let X' be a subdivision of a polyhedral complex X. Then the set of good functions for the subdivision X' form an open convex polyhedral cone in PL(X').

Corollary 1.10. Let X be a polyhedral complex with a rational structure M and let X' be a rational, projective subdivision of X. Then we can find a good function f for the subdivision X' such that for each polyhedron τ in X', $f|_\tau \in M_\tau$.

Proof. Let $C \subset PL(X')$ be the open cone of good functions. By assumption $C \neq \emptyset$.

Since X' is a rational subdivision it carries a rational structure and therefore PL(X') considered as a subspace of $\mathbb{R}^{X_0'}$ is defined over the rationals, i.e.,

$$PL(X') \cap \mathbb{Q}^{X_0'} \text{ is dense in } PL(X').$$

Let g be any element of $C \cap \mathbb{Q}^{X_0'}$. Then a suitable multiple $f = n.g$ will do. q.e.d.

The following follows from the proof of Lemma 1.8:

Lemma 1.11. Let X, X', X'' be three polyhedral complexes such that X' is a subdivision of X and X'' is a subdivision of X'. Then we can find two homogeneous convex functions Δ' and Δ'' on $PL(X'')/PL(X)$ and $PL(X'')/PL(X')$ respectively such that:

$f \in PL(X')$ is a good function for the subdivision X' of X $\iff \Delta'(f) > 0$

$g \in PL(X'')$ is a good function for the subdivision X'' of X' $\iff \Delta''(g) > 0$

$h \in PL(X'')$ is a good function for the subdivision X'' of X $\iff \min\{\Delta'(h), \Delta''(h)\} > 0$

Corollary 1.12. (Transitivity of projective subdivisions) Let X, X', and X'' be polyhedral complexes such that X' is a projective subdivision of X and X'' is a projective subdivision of X'. Then X'' is a projective subdivision of X.

Proof. Let Δ' and Δ'' be as in Lemma 2.9. By assumption we can find functions f in $PL(X')$, g in $PL(X'')$ such that

$$\Delta'(f) = \Theta_1 > 0$$

$$|\Delta'(g)| = K < \infty$$

$$\Delta''(f) = 0$$

$$\Delta''(g) = \Theta_2 > 0 .$$

Let $\epsilon > 0$ be any real number such that $\Theta_1 - \epsilon K > 0$. Then

$$\min\Big(\Delta'(f+\epsilon g),\ \Delta''(f+\epsilon g)\Big) \geq \min\Big(\Theta_1 - \epsilon K,\ \epsilon\Theta_2\Big) > 0 .$$

Hence $f+\epsilon g$ is a good function for the subdivision X" of X.

q.e.d.

§2. Some examples of projective subdivisions

§2A

Example 2.1. (The barycentric subdivision)

Let X be a polyhedral complex of dimension n and let X' be the barycentric subdivision.

Let $X^{(k)}$, $0 \leq k \leq n-1$, be the subdivision of X obtained by taking the cones over the barycenters of the n-polyhedrons, n-1-polyhedrons,$\cdots$,n-k-polyhedrons taken in this order. We then have a succession of subdivisions.

$$X = X^{(0)}, X^{(1)}, \cdots, X^{(k)}, \cdots, X^{(n-1)} = X'.$$

Let f_k be the function which has value 1 on all the barycenters of the n-k-polyhedra in X, 0 on all the vertexes of $X^{(k-1)}$ and linear on each polyhedron in $X^{(k)}$. Then clearly f_k is a good function for the subdivision $X^{(k)}$ of $X^{(k-1)}$. By the transitivity of projective subdivisions we get

Lemma 2.2. The barycentric subdivision is projective.

§2B

Example 2.3 (The regular subdivision of a simplex).

We will consider a simplex Δ with an ordered set of vertexes $P_1, \cdots, P_{n+1}$.

If $x_1,\cdots,x_{n+1}$ are the barycentric coordinates, we define the cumulative coordinates as follows.

$$
\begin{aligned}
y_o &= 0\\
y_1 &= x_1\\
y_2 &= x_1+x_2\\
&\vdots\\
y_k &= x_1 + x_2 +\cdots+ x_k\\
&\vdots\\
y_{n+1} &= 1
\end{aligned}
$$

We then identify Δ with the set of n-tuples of real numbers $y_1,\cdots,y_n$ such that

$$0 \leq y_1 \leq y_2 \leq \cdots \leq y_n \leq 1.$$

Let μ be a positive integer and consider the hyperplanes

$$H_k^{i,j}: \quad y_i-y_j = k/\mu$$

where $0 \leq j < i \leq n$; $0 \leq k \leq \mu$.

Lemma 2.4. The hyperplanes H_k^{ij} define a subdivision of Δ which we will call $\Delta^{(\mu)}$. $\Delta^{(\mu)}$ is a simplicial complex and if σ is any n-simplex of $\Delta^{(\mu)}$ then volume $\sigma = 1/\mu^n \cdot$volume Δ.

Proof. Let $y_1,\cdots,y_n$ be a point in $\Delta' = \text{int}\,\Delta - \bigcup H_k^{ij}$.

If we put $a_i = [\mu\cdot y_i]$ and $t_i = y_i - a_i/\mu$ it is clear that $t_i \neq t_j$ whenever $i \neq j$. Hence there is a unique permutation σ such that:

$$0 < t_{\sigma(1)} < t_{\sigma(2)} < \cdots < t_{\sigma(n)} < 1/\mu .$$

Also it is clear that the connected component of Δ' determined by the point $y_1 \cdots y_n$ is the set of points $y_1' \cdots y_n'$ such that

$$0 < y'_{\sigma(1)} - a_{\sigma(1)}/\mu < \cdots < y'_{\sigma(n)} - a_{\sigma(n)}/\mu < \frac{1}{\mu} .$$

And this we see is the interior of a simplex which modulo a permutation of the coordinate axes and a translation equals $1/\mu \cdot \Delta$, and hence its volume is $1/\mu^n \cdot$volume Δ.

q.e.d.

The analogous subdivision of the entire $\mathbb{R}^n$ by the hyperplanes H_k^{ij} (all $k \in \mathbb{Z}$) is just the decomposition into affine Weyl chambers of type A_n.

<u>Lemma</u> 2.4': <u>The regular subdivision $\Delta^{(\mu)}$ is projective.</u>

<u>Proof</u>. Let f be the function

$$f = \sum_{\substack{i_1, i_2, k \\ 1 \leq i_1 < i_2 \leq n+1 \\ 1 \leq k \leq \mu-1}} \left| \sum_{i_1 \leq i \leq i_2} x_i - \frac{k}{\mu} \right| .$$

It is easy to see that f is a good function.

In order to treat the more difficult mixed (μ,ν)-subdivision discussed below, we need to analyze the regular subdivision in much greater detail, describing explicitly all its simplices, their vertices, and their faces. This will occupy us through (2.17) below.

Lemma 2.5. *There is a natural 1-1 correspondence between n-simplexes of $\Delta^{(\mu)}$ and maps $\pi: \{1,\cdots,n\} \longrightarrow \{0,\cdots,\mu-1\}$.*

Proof. Suppose first we are given a simplex in $\Delta^{(\mu)}$. Then as in the proof of the last lemma, for any interior point $y_1\cdots y_n$ we get the numbers $a_i = [\mu\cdot y_i]$ and a permutation σ. This does not depend on the interior point chosen. The corresponding map will then be $i \leadsto a_{\sigma(i)}$.

Conversely, given a map $\pi: \{1,\cdots,n\} \longrightarrow \{0,\cdots,\mu-1\}$ there is a unique permutation σ satisfying:

i) $\pi\sigma(i) \leq \pi\sigma(j)$ for $i \leq j$

ii) $\pi\sigma(i) = \pi\sigma(j) \Longrightarrow (i < j \Longleftrightarrow \sigma(i) < \sigma(j))$.

In fact we have the formula

$$\underline{\sigma^{-1}(k) = \#\left\{i \in \{1,\cdots,n\} \,\middle|\, \pi(i) < \pi(k) \text{ or } i \leq k \text{ and } \pi(i) = \pi(k)\right\}}.$$

If $0 \leq t_{\sigma^{-1}(1)} \leq t_{\sigma^{-1}(2)} \leq \cdots \leq t_{\sigma^{-1}(n)} \leq \frac{1}{\mu}$, it follows by i) and ii) that the point $y_1,\cdots,y_n$ defined by

$$y_i = \frac{\pi\sigma(i)}{\mu} + t_i$$

lies in Δ. By varying the t's we get a simplex of $\Delta^{(\mu)}$ which we denote by Δ^{π}.

It is immediate from these definitions that the two maps are inverse to each other. q.e.d.

Definition 2.6. Let Δ, μ, π, σ be as in the last lemma. We define

$$\delta_k(i) = \begin{cases} 1/\mu & \text{if } \sigma(i) \geq k \\ 0 & \text{if } \sigma(i) < k \end{cases}$$

$\forall\ 1 \leq i \leq n, \quad \forall\ 1 \leq k \leq n+1.$

Clearly we have:

$$0 \leq \delta_k(\sigma^{-1}(1)) \leq \delta_k(\sigma^{-1}(2)) \leq \cdots \leq \delta_k(\sigma^{-1}(n)) \leq \frac{1}{\mu} .$$

Hence the point P_k^{π} given by

$$y_i(P_k^{\pi}) = \frac{\pi\sigma(i)}{\mu} + \delta_k(i)$$

lies in Δ^{π}. We call this particular ordering of the vertexes of Δ^{π} the canonical ordering.

Example of regular subdivision in the case $n = 2$, $\mu = 3$:

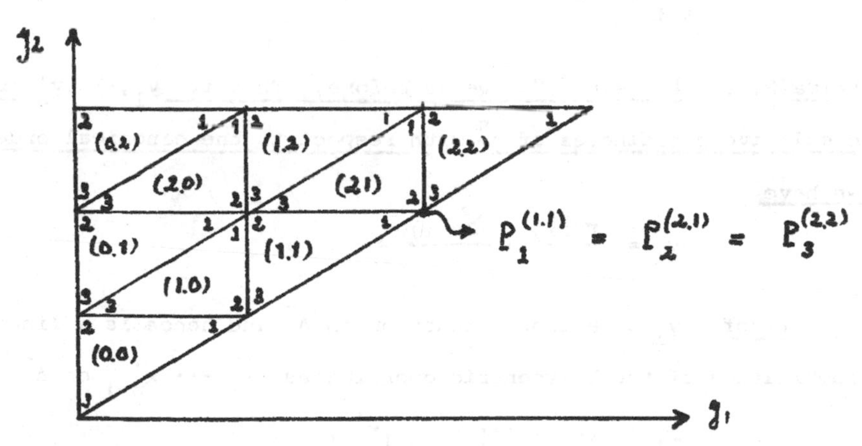

In each triangle we have indicated the function π and the induced ordering of the vertices.

<u>Lemma</u> 2.7. <u>Let Δ,μ,π,σ as before, and let $\alpha_1,\alpha_2,\cdots,\alpha_n$ be real negative numbers which are linearly independent over the rationals. The linear function $\ell = \sum \alpha_i y_i$ separates all the vertexes of $\Delta^{(\mu)}$ and hence defines a total ordering on the vertexes of $\Delta^{(\mu)}$. Moreover this ordering induces the canonical ordering on each simplex of $\Delta^{(\mu)}$.</u>

<u>Proof</u>. Let P_j^π and P_k^π be two vertexes of Δ^π such that $j < k$. We then have

$$\ell(P_j^\pi) - \ell(P_k^\pi) =$$

$$\sum_i \alpha_i \frac{\pi\sigma(i)}{\mu} + \alpha_i \delta_j(i) - \alpha_i \frac{\pi\sigma(i)}{\mu} - \alpha_i \delta_k(i) =$$

$$\sum_i \alpha_i(\delta_j(i)-\delta_k(i)) =$$

$$\sum_i \alpha_{\sigma^{-1}(i)}(\delta_j(\sigma^{-1}(i)) - \delta_k(\sigma^{-1}(i)) =$$

$$\sum_{i=j}^{k-1} \alpha_{\sigma^{-1}(i)}/\mu < 0 .$$

q.e.d.

<u>Lemma</u> 2.8. <u>Let Δ,μ,π,σ be as before. Then if $y_1',\cdots,y_n'$ are the cumulative coordinates of Δ^π with respect to the canonical ordering we have</u>

$$y_i = \frac{\pi\sigma(i)}{\mu} + \frac{y'_{\sigma(i)}}{\mu}$$

<u>Proof</u>. y_i is a linear function on Δ^π and hence is a linear combination of the barycentric coordinates $x_1',\cdots,x_{n+1}'$ on Δ^π. Say

$$y_i = \alpha_1 x_1' + \cdots + \alpha_{n+1} x_{n+1}' .$$

But the coefficients are given by

$$\alpha_k = y_i(P_k^{\pi}) = \frac{\pi\sigma(i)}{\mu} + \delta_k(i)$$

and so

$$y_i = \sum_{k=1}^{n+1} x_k' \cdot \frac{\pi\sigma(i)}{\mu} + \sum_{k=1}^{n+1} x_k' \cdot \delta_k(i)$$

$$= \frac{\pi\sigma(i)}{\mu} + \sum_{k\leq\sigma(i)} \frac{x_k'}{\mu} = \frac{\pi\sigma(i)}{\mu} + \frac{y'_{\sigma(i)}}{\mu} \ . \qquad \text{q.e.d.}$$

Corollary 2.9. *Let ν be a positive integer and subdivide each simplex of $\Delta^{(\mu)}$ with respect to ν and the canonical ordering. By Lemma 2.7 these subdivisions patch up and give us a subdivision of Δ which we denote by $\Delta^{(\mu)(\nu)}$. We have*

$$\Delta^{(\mu)(\nu)} = \Delta^{(\mu\cdot\nu)} \ .$$

Proof. Immediate by the formula of Lemma 2.8. q.e.d.

Next we want to study some particular properties of adjacent n-simplexes in $\Delta^{(\mu)}$. Let π be a map $\{1,\cdots,n\} \longrightarrow \{0,\cdots,\mu-1\}$, and σ the associated permutation. We may consider σ as a permutation of the set $\{0,1,\cdots,n+1\}$ leaving 0 and $n+1$ fixed. If $P_1^{\pi},\cdots,P_{n+1}^{\pi}$ are the vertexes of Δ^{π} taken with the canonical ordering, the equation of the plane which goes through $P_1^{\pi},\cdots,\hat{P}_k^{\pi},\cdots,P_{n+1}^{\pi}$ is:

$$y_k' - y_{k-1}' = 0 \ .$$

Here the convention is $y'_o \equiv 0$, $y'_{n+1} \equiv 1$. By Lemma 2.8 this equation is

$$y_{\sigma^{-1}(k)} - y_{\sigma^{-1}(k-1)} = \frac{1}{\mu}(\pi(k-\pi(k-1))) .$$

Here the convention is $\pi(o) \equiv 0$, $\pi(n+1) \equiv \mu-1$.

<u>Lemma</u> 2.10. <u>If we denote the plane through the vertexes</u> $P_1^\pi, P_2^\pi, \cdots, P_k^\pi, \cdots, P_{n+1}^\pi$ <u>by</u> H_k^π, $1 \leq k \leq$ <u>n+1, we have</u>

$$H_k^\pi \text{ is a face of } \Delta \Longleftrightarrow \pi(k) = \pi(k-1).$$

<u>Proof</u>. The equation of H_k^π is

$$y_{\sigma^{-1}(k)} - y_{\sigma^{-1}(k-1)} = \frac{1}{\mu}(\pi(k)-\pi(k-1)) .$$

Since the equations of the faces of Δ are given by equations

$$y_i - y_{i-1} = 0$$

the if part is clear.

Conversely if $\pi(k) - \pi(k-1) = 0$

$$\sigma^{-1}(k) = \# \left\{ i \in \{1,\cdots,n\} \middle| \pi(i) < \pi(k) \quad \cup \quad i \in \{1\cdots n\} \middle| \pi(i)=\pi(k),\ i \leq k \right\}$$

$$\sigma^{-1}(k-1) = \# \left\{ i \in \{1\cdots n\} \middle| \pi(i) < \pi(k-1) \quad \cup \quad i \in \{1\cdots n\} \middle| \pi(i)=\pi(k-1),\ i < k-1 \right\}$$

We see that the first set is the second set with the element k adjoined so

$$\sigma^{-1}(k) = \sigma^{-1}(k-1) + 1 .$$

In the next lemma we intend to write down the coordinates of the vertices of two adjacent simplexes. The reader is advised to work out a couple of low-dimensional examples instead of reading through this mess.

Let Δ,μ,π,σ be as before and let k be an integer such that $2 \leq k \leq n$. Suppose moreover that $\pi(k) \neq \pi(k-1)$. By the last lemma this means that there is a simplex adjacent to Δ^{π} opposite the vertex P_k^{π}.

Let ϵ_k be the permutation of $\{0,\cdots,n+1\}$ obtained by interchanging k and k-1, i.e.,

$$\epsilon_k(i) = i \qquad \text{for} \qquad i \notin \{k,k-1\}$$

$$\epsilon_k(k-1) = k$$

$$\epsilon_k(k) = k-1.$$

Let π' be the function $\pi\cdot\epsilon_k$, and let σ' be the permutation $\epsilon_k\cdot\sigma$. Then

Lemma 2.11. <u>With the above notation, $\Delta^{\pi'}$ is the simplex adjacent to Δ^{π} opposite the vertex P_k^{π}. σ' is the permutation associated to π'.</u>

<u>Proof</u>. First we show that σ' is the right permutation, i.e., has the property i) and ii) of Lemma 2.5. i) is immediate since $\pi'\sigma' = \pi\epsilon_k\epsilon_k\sigma = \pi\sigma$. Suppose then that $\pi'\sigma'(i) = \pi'\sigma'(j)$. But then $\pi\sigma(i) = \pi\sigma(j)$ so we have $i < j \iff \sigma(i) < \sigma(j)$. By our assumption $\pi(k) \neq \pi(k-1)$ hence the pair $\{\sigma(i),\sigma(j)\}$ is not equal the pair

$\{k,k-1\}$, so $\sigma(i) < \sigma(j) \Longleftrightarrow \sigma'(j) < \sigma'(j)$.

To show that Δ^{π} and $\Delta^{\pi'}$ are adjacent we compute the coordinates of the vertexes.

Let δ be the function defined in Definition 2.6, and let δ' be the corresponding function for σ'. The vertexes are then given by:

$$y_i(P_j^{\pi}) = \frac{\pi\sigma(i)}{\mu} + \delta_j(i)$$

$$y_i(P_j^{\pi'}) = \frac{\pi'\sigma'(i)}{\mu} + \delta_j'(i) .$$

Since $\pi\sigma = \pi'\sigma'$ we just have to compare δ and δ'. But if $j \neq k$ we have $\epsilon_k(i) \geq j \Longleftrightarrow i \geq j$ and so $\delta_j(i) = \delta_j(i)$ for all i. It follows that $P_j^{\pi} = P_j^{\pi'}$ for $j \neq k$.

q.e.d.

Lemma 2.12 (Quadrilateral lemma in the case $2 \leq k \leq n$).

The four points $P_{k-1}^{\pi}, P_k^{\pi}, P_{k+1}^{\pi}, P_k^{\pi'}$ lie in a plane and the two line segments $P_k^{\pi}P_k^{\pi'}$ and $P_{k-1}^{\pi}P_{k+1}^{\pi}$ cut each other in half.

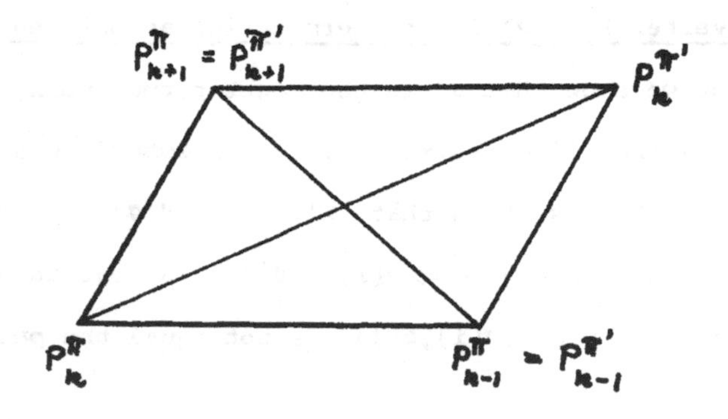

Proof. Clearly $\delta_{k-1}(i) = \delta_k(i) = \delta'_k(i) = \delta_{k+1}(i)$ for $i \notin \{\sigma^{-1}(k-1), \sigma^{-1}(k), \sigma^{-1}(k+1)\}$.

When $i = \sigma^{-1}(k-1)$ we have

$$\delta_{k-1}(i) = 1/\mu,\ \delta_k(i) = 0,\ \delta'_k(i) = 1/\mu,\ \delta_{k+1}(i) = 0.$$

When $i = \sigma^{-1}(k)$ we have

$$\delta_{k-1}(i) = 1/\mu,\ \delta_k(i) = 1/\mu,\ \delta'_k(i) = 0,\ \delta_{k+1}(i) = 0.$$

When $i = \sigma^{-1}(k+1)$ we have

$$\delta_{k-1}(i) = 1/\mu,\ \delta_k(i) = 1/\mu,\ \delta'_k(i) = 1/\mu,\ \delta_{k+1}(i) = 1/\mu.$$

In all cases

$$\delta_{k-1}(i) + \delta_{k+1}(i) = \delta_k(i) + \delta'_k(i)\ .$$

Hence for all i

$$y_i(P^{\pi}_{k-1}) + y_i(P^{\pi}_{k+1}) = y_i(P^{\pi}_k) + y_i(P^{\pi'}_k). \qquad \text{q.e.d.}$$

Next we consider the case $k = 1$ and we assume that $\pi(1) \neq \pi(0) = 0$. By Lemma 2.10, there is a simplex adjacent to Δ^{π} opposite the vertex P^{π}_1.

Let π' be the function $\{0,\cdots,n+1\} \longrightarrow \{0,\cdots,\mu\}$ defined by $\pi'(0) = 0$, $\pi'(n+1) = \mu-1$, $\pi'(n) = \pi(1)-1$ and $\pi'(i) = \pi(i+1)$ for $i \notin \{0,n,n+1\}$.

Lemma 2.13. $\Delta^{\pi'}$ is the simplex adjacent to Δ^{π} opposite the vertex P_1^{π}.

Proof. The proof is pure calculation. We calculate the coordinate functions of Δ^{π} and $\Delta^{\pi'}$ and compare. Recall that σ' is given by the formula

$$\sigma'^{-1}(k) = \#\left\{i\in\{1\cdots n\}\,\middle|\,\pi'(i)<\pi'(k)\right\} + \#\left\{i\in\{1,\cdots,k\}\,\middle|\,\pi'(i) = \pi'(k)\right\}.$$

Hence $\sigma'^{-1}(0) = 0, \quad \sigma'^{-1}(n+1) = n+1$

$$\begin{aligned}\sigma'^{-1}(n) &= \#\left\{i\in\{1\cdots n\}\,\middle|\,\pi'(i)<\pi(1)-1\right\} + \#\left\{i\in\{1\cdots n\}\,\middle|\,\pi'(i) = \pi(1)-1\right\}\\ &= \#\left\{i \in \{1\cdots n\}\,\middle|\,\pi'(i) < \pi(1)\right\}\\ &= \#\left\{i \in \{1\cdots n-1\}\,\middle|\,\pi(i+1) < \pi(1)\right\} + 1\\ &= \#\left\{i \in \{1\cdots n\}\,\middle|\,\pi(i)<\pi(1)\right\} + \#\left\{i\in\{1\}\,\middle|\,\pi(i) = \pi(1)\right\}\\ &= \underline{\sigma^{-1}(1)}.\end{aligned}$$

For $k \in \{1,\cdots,n-1\}$ we have

$$\begin{aligned}\sigma'^{-1}(k) &= \#\left\{i\in\{1\cdots n\}\,\middle|\,\pi'(i)<\pi(k+1)\right\} + \#\left\{i\in\{1\cdots k\}\,\middle|\,\pi(i+1) = \pi(k+1)\right\}\\ &= \#\left\{i\in\{1\cdots n\}\,\middle|\,\pi(i) < \pi(k+1)\right\} + \delta_{\pi(1),\pi(k+1)}\\ &\quad + \#\left\{i\in\{1\cdots k+1\}\,\middle|\,\pi(i) = \pi(k+1)\right\} - \delta_{\pi(1),\pi(k+1)}\\ &= \underline{\sigma^{-1}(k+1)}\,.\end{aligned}$$

Let ϵ be the permutation defined by

$\epsilon(0) = 0, \quad \epsilon(n+1) = n+1, \quad \epsilon(n) = 1, \quad \epsilon(i) = i+1 \text{ if } i \in \{1,\cdots,n-1\}.$

We then have

$$\underline{\pi'(i) = \pi \cdot \epsilon(i) - \delta_{n,i}}$$

$$\underline{\sigma'(i) = \epsilon^{-1} \cdot \sigma(i)}$$

$$\underline{\pi'\sigma(i)} = \pi \in (\epsilon^{-1}(\sigma(i))) - \delta_{n,\epsilon^{-1}\sigma(i)}$$

$$= \pi\sigma(i) - \delta_{\epsilon(n),\sigma(i)} = \underline{\pi\sigma(i)-\delta_{1,\sigma(i)}} \; .$$

As before, we define the functions

$$\delta_i(k) = \begin{Bmatrix} \frac{1}{\mu} \Longleftrightarrow \sigma(k) \geq i \\ 0 \Longleftrightarrow \sigma(k) < i \end{Bmatrix}$$

$$\delta_i'(k) = \begin{Bmatrix} \frac{1}{\mu} \Longleftrightarrow \sigma'(k) \geq i \\ 0 \Longleftrightarrow \sigma'(k) < i \end{Bmatrix}$$

Let $i \in \{1,\cdots,n\}$, and consider the two vertexes P^{π}_{i+1} and $P^{\pi'}_{i}$. Coordinates are given by

$$y_k(P^{\pi}_{i+1}) = \frac{\pi\sigma(k)}{\mu} + \delta_{i+1}(k)$$

$$y_k(P^{\pi'}_{i}) = \frac{\pi'\sigma'(k)}{\mu} + \delta_i'(k).$$

<u>Case 1</u>. $k \neq \sigma^{-1}(1)$.

In this case we have $\pi\sigma(k) = \pi'\sigma'(k)$. Moreover: $k = \sigma^{-1}(1) \Longleftrightarrow \sigma'(k) = n$, so

$$\delta_{i+1}(k) = \frac{1}{\mu} \Longleftrightarrow \sigma(k) \geq i+1 \Longleftrightarrow \epsilon\sigma'(k) \geq i+1$$

$$\Longleftrightarrow \sigma'(k) \geq i \quad (\text{since } \sigma'(k) \neq n) \Longleftrightarrow \delta_i'(k) = \frac{1}{\mu} \; .$$

Hence P^{π}_{i+1} and $P^{\pi'}_{i}$ have the same y_k-coordinates for $k \neq \sigma^{-1}(1)$.

<u>Case 2</u>. $k = \sigma^{-1}(1) \Longleftrightarrow \sigma'(k) = n$.

We have $\sigma(k) = 1 < i+1$ for all i

and $\sigma'(k) = n \geq i$ for all i.

Hence:

$$y_k(P^{\pi}_{i+1}) = \frac{\pi\sigma(k)}{\mu}$$

$$y_k(P^{\pi'}_{i}) = \frac{\pi'\sigma'(k)}{\mu} + \frac{1}{\mu}$$

$$= \frac{\pi\sigma(k)}{\mu} - \frac{\delta_{1,\sigma(k)}}{\mu} + \frac{1}{\mu} = \frac{\pi\sigma(k)}{\mu}.$$

This shows that $P^{\pi}_{i+1} = P^{\pi'}_{i}$ for $i \in \{1 \cdots n\}$. q.e.d.

<u>Lemma 2.14.</u> <u>(Quadrilateral lemma in the case k = 1).</u>

<u>With the notation as in Lemma 2.13 the four points $P^{\pi}_1, P^{\pi'}_{n+1}, P^{\pi}_2$ and P^{π}_{n+1} lie in a plane and the two line segments $P^{\pi}_1 P^{\pi'}_{n+1}$ and $P^{\pi}_2 P^{\pi}_{n+1}$ cut each other in half.</u>

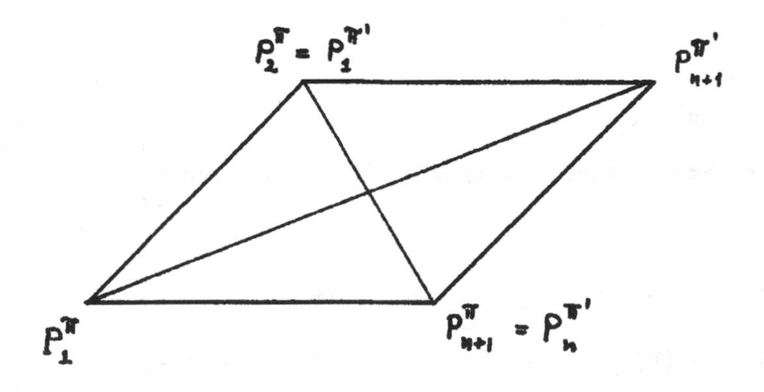

<u>Proof</u>. $y_k(P_1^{\pi}) + y_k(P_{n+1}^{\pi'}) = \frac{\pi\sigma(k)}{\mu} + \delta_1(k) + \frac{\pi'\sigma'(k)}{\mu} + \delta'_{n+1}(k)$

$$= \frac{\pi\sigma(k)}{\mu} + \frac{\pi'\sigma'(k)}{\mu} + \frac{1}{\mu}$$

$$y_k(P_2^{\pi}) + y_k(P_{n+1}^{\pi}) = y_k(P_1^{\pi'}) + y_k(P_{n+1}^{\pi})$$

$$= \frac{\pi'\sigma'(k)}{\mu} + \delta'_1(k) + \frac{\pi\sigma(k)}{\mu} + \delta_{n+1}(k)$$

$$= \frac{\pi\sigma(k)}{\mu} + \frac{\pi'\sigma'(k)}{\mu} + \frac{1}{\mu}. \qquad \text{q.e.d.}$$

Note that the case $k = n+1$ follows from this case by interchanging π and π'.

<u>Definition</u> 2.15. <u>(Good and bad hyperplanes)</u>

<u>Let $\Delta = \{P_1,\cdots,P_{n+1}\}$ be an n-simplex with an ordered set of vertexes</u>, $y_1,\cdots,y_n$ <u>the cumulative coordinates and μ a positive integer.</u>

<u>A hyperplane $H_k^{i,j}$ defined by the equation</u>

$$y_j - y_i = k/\mu, \qquad 0 \le i < j \le n+1$$

<u>is said to be a good hyperplane if it can be written in the form</u>

$$y_{\ell} = k'/\mu, \qquad 1 \le \ell \le n .$$

<u>Note that this is the case if and only if $j = n+1$ or $i = 0$.</u>

<u>The hyperplanes which are not of this form will be called bad.</u>

We now state a souped-up version of the last 5 lemmas.

Lemma 2.16. Let again Δ be an n-simplex with ordered set of vertexes $\{P_1 \cdots P_{n+1}\}$, and let σ and τ be two adjacent n-simplexes in the μ-regular subdivision of Δ. Let H be the hyperplane that separates σ and τ. If P is the vertex of σ which does not meet τ and Q is the vertex of τ that does not meet σ, there are two vertexes A and B in $\sigma \cap \tau$ such that the four points P, Q, A and B lie in a plane and the line segments PQ and AB cut each other in half.

Moreover if x_{n+1} is the $n+1^{st}$ barycentric coordinate and $x_{n+1}(P) = x_{n+1}(Q)$ then $x_{n+1}(A) = x_{n+1}(B)$ and the plane A,B,P,Q lies in the hyperplane x_{n+1} = constant.

However if $x_{n+1}(P) \neq x_{n+1}(Q)$, then $x_{n+1}(A) \neq x_{n+1}(B)$ and the hyperplane H cuts the face opposite P_{n+1} in a good hyperplane, or H is defined by x_{n+1} = constant.

Proof. The proof of the last two assertions follows directly from the formulas developed so far and is left to the reader.

Observation 2.17. Let Δ be an n-simplex on our ordered set of vertexes $P_1 \cdots P_{n+1}$, μ an integer > 0 and k any integer $0 \leq k < \mu$.

Let H be the hyperplane defined by $x_{n+1} = k/\mu$, and let P_i' be the intersection point of H with the line P_i, P_{n+1}, $1 \leq i \leq n$.

We denote by Δ' the truncated simplex with vertexes $\{P_1' \cdots P_n', P_{n+1}\}$. In this case we have:

$$\Delta' \cap \Delta^{(\mu)} = \Delta'^{(\mu-k)}$$

i.e., the μ-regular subdivision of Δ induces the $(\mu-k)$-regular subdivision of Δ'.

Proof. Left to reader.

§2C

<u>Example</u> 2.18. (<u>The mixed</u> (ν,μ) <u>subdivision</u>)

We consider a simplex Δ with vertexes $\{P_1,\cdots,P_{n+1},Q\}$ taken in this order, and we denote the face opposite Q by $\Delta_Q = \{P_1,\cdots,P_{n+1}\}$.

If $\pi: \{1,\cdots n\} \longrightarrow \{0,\cdots \nu-1\}$ is any function we denote as before by Δ_Q^{π} the corresponding simplex in $\Delta_Q^{(\nu)}$.

Let Δ^{π} be the simplex $\{P_1^{\pi},\cdots,P_{n+1}^{\pi},Q\}$ with vertexes taken in this order:

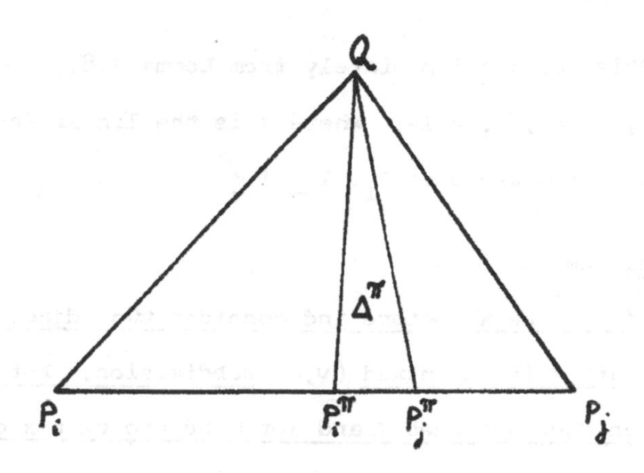

If we now subdivide each Δ^{π} regularly with respect to μ and the given ordering, these subdivisions clearly patch up so as to give a subdivision of Δ. We call this the mixed (ν,μ)-subdivision of Δ with respect to this ordering. It is easy to see using 1.12 and 2.4' that this subdivision is projective. The obvious good functions would not however extend to the global Example D which is our main goal so we do not stop to do this. Instead we want some facts about adjacent simplices in this subdivision ((2.20)) which are the key points for establishing projectivity in the general case.

Lemma 2.19. Let Δ, π, μ, ν be as above. If $y_1, \cdots, y_{n+1}$ are the cumulative coordinates on Δ and $y_1^\pi, \cdots, y_{n+1}^\pi$ are the cumulative coordinates on Δ^π we have the formula

$$y_k = \frac{1}{\nu}(\pi\sigma(k) \cdot y_{n+1}^\pi + y_{\sigma(k)}^\pi \qquad 1 \leq k \leq n+1$$

$$y_{n+1} = y_{n+1}^\pi .$$

Proof. This follows immediately from Lemma 2.8.

Note that $y_{n+1} = y_{n+1}^\pi = 1-\ell$ where ℓ is the linear function which takes the value 1 at Q and 0 at P_i, $1 \leq i \leq n$. q.e.d.

Lemma 2.20 (Main Lemma)

Let Δ and $\Delta^{(\nu,\mu)}$ be as before and consider two adjacent n+1-simplexes σ and τ in the mixed (ν,μ)-subdivision. Let P be the vertex of σ which does not meet τ and let Q be the vertex of τ that does not meet σ. Let H be the hyperplane that separates σ and τ. Now three things may happen.

a) The hyperplane H is defined by the equation ℓ = constant

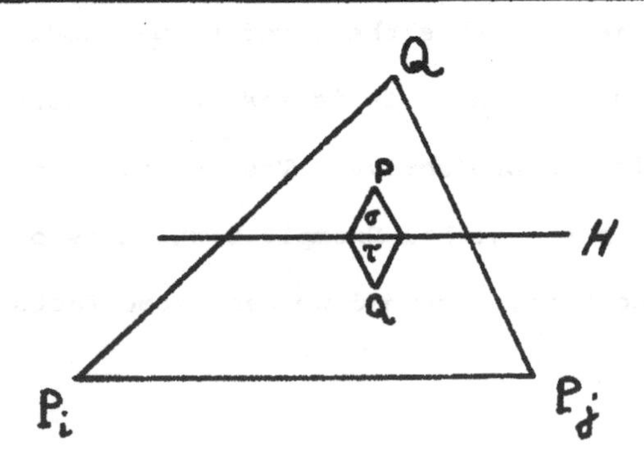

b) The hyperplane H instersects $\Delta_Q = \{P_1, \cdots, P_{n+1}\}$ in a good hyperplane.

In this case we may assume that $\ell(P) = k/\mu$ and $\ell(Q) = k-1/\mu$, $1 \leq k \leq \mu-1$. And the important fact is that the part of the hyperplane H which lies below $\ell = k/\mu$ does not meet the interior of any n+1-simplex in the subdivision.

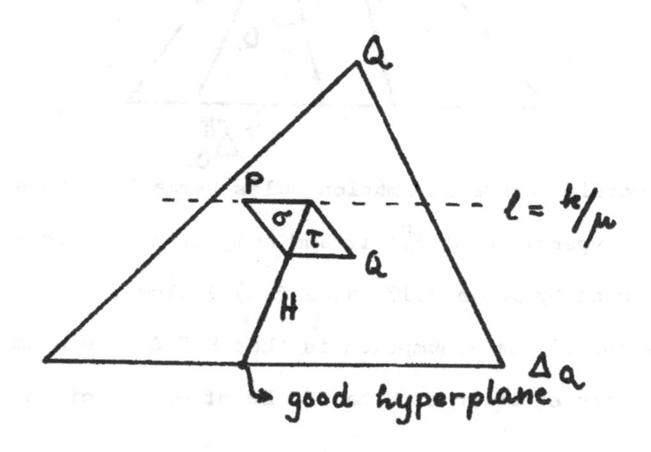

c) The hyperplane H intersects Δ_Q in a bad hyperplane.

[In this case H might cut through interiors of n+1-simplexes in the subdivision even in the level below $\ell = \max(\ell(Q), \ell(P))$. Hence the name bad.]

In this case however the following fortunate thing happens: There are two vertexes A and B of $\sigma \cap \tau$ such that the four points P,Q,A and B lie in a plane, the two line segments PQ and AB cut each other in half and moreover $\ell(P) = \ell(Q) = \ell(A) = \ell(B)$, i.e., the four points lie in the hyperplane defined by ℓ = constant.

Proof. We first consider the case where σ and τ belong to the same simplex Δ^{π}:

i.e.

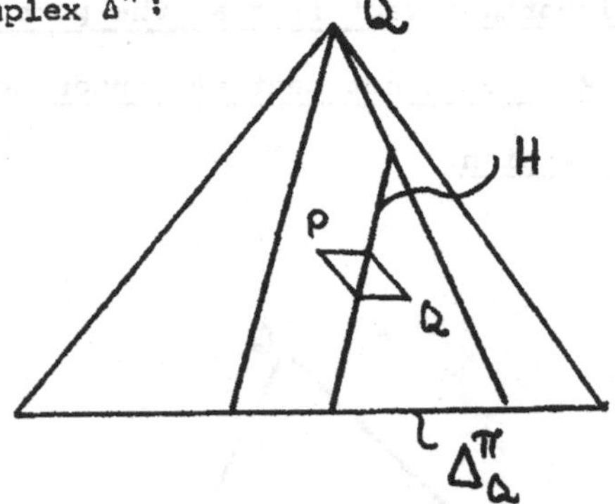

By the coordinate transformation rules Lemma 2.8, we see that $H \cap \Delta^{\pi}_{Q}$ is a good hyperplane of Δ^{π}_{Q} if and only if $H \cap \Delta_{Q}$ is a good hyperplane of Δ_{Q}, and so by Lemma 2.17 a) and c) follows.

Now for b) the assumption is that $H \cap \Delta^{\pi}_{Q}$ is a good hyperplane. Therefore the equation of H has to be of either of the following two forms.

(i) $$y^{\pi}_{n+1} - y^{\pi}_{j} = \frac{\mu-k}{\mu}$$

(ii) $$y^{\pi}_{j} = \frac{k}{\mu} \qquad 1 \leq k \leq \mu-1 .$$

Since $y^{\pi}_{n+1} = 1-\ell$ restricted to Δ^{π}_{Q} is identically equal to 1, both of these equations restrict to the equation

$$y^{\pi}_{j} = \frac{k}{\mu} \text{ on } \Delta^{\pi}_{Q} .$$

Using the coordinates transformation formula in Lemma 2.19 we get H defined by the following equation.

(i) $$y_{\sigma^{-1}(j)} - \frac{\pi(j)+1}{\nu}(1-\ell) = \frac{k-\mu}{\mu\cdot\nu}$$

(ii) $$y_{\sigma^{-1}(j)} - \frac{\pi(j)}{\nu}(1-\ell) = \frac{k}{\mu\cdot\nu}$$

Since $\ell = 0$ on Δ_Q both equations restrict to the equation

$$y_{\sigma^{-1}(j)} = \frac{k}{\mu\cdot\nu} + \frac{\pi(j)}{\nu} .$$

Consider the two hyperplanes K_1 and K_2 defined by

$$K_1: \quad y_{\sigma^{-1}(j)} - \frac{\pi(j)}{\nu}(1-\ell) = 0$$

$$K_2: \quad y_{\sigma^{-1}(j)} - \frac{\pi(j)+1}{\nu}(1-\ell) = 0$$

We have the following self explanatory picture:

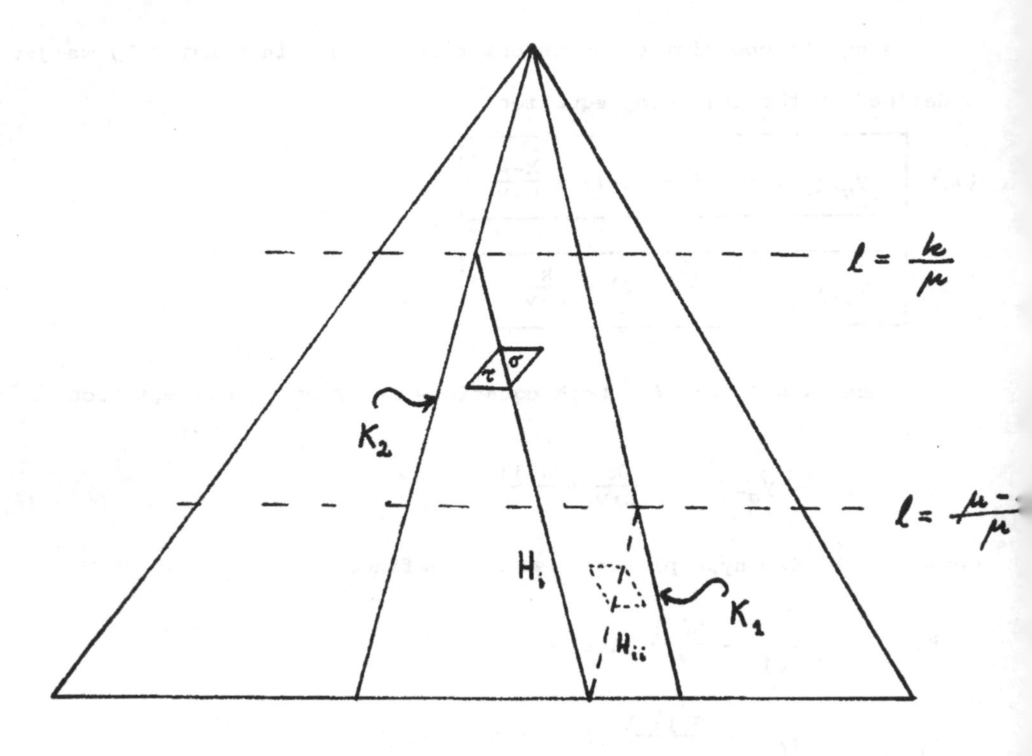

b) will follow if we can prove that for any simplex of the form $\Delta^{\pi'}$ which is trapped between K_1 and K_2 the hyperplane $H \cap \Delta^{\pi'}$ is one of the hyperplanes which defines the μ-regular subdivision of $\Delta^{\pi'}$, i.e., the equation of H in the accumulative π'-coordinates is of the form

$$y_j^{\pi'} - y_i^{\pi'} = \frac{k}{\mu} .$$

Now the condition that $\Delta^{\pi'}$ is trapped between K_1 and K_2 is the same as that of the simplex $\Delta_Q^{\pi'}$ lying betwen $K_1 \cap \Delta_Q$ and $K_2 \cap \Delta_Q$. Hence for each vertex $P_i^{\pi'}$ of $\Delta_Q^{\pi'}$ we must have

$$\frac{\pi(j)}{\nu} \leq y_{\sigma^{-1}(j)}(P_i^{\pi'}) \leq \frac{\pi(j)+1}{\nu} .$$

But by some old formula we have

$$y_{\sigma^{-1}(j)}(P_i^{\pi'}) = \frac{\pi'\sigma'(\sigma^{-1}(j))}{\nu} + \delta_i(\sigma^{-1}(j)) .$$

Since this holds for all i the trapping condition reads

$$\pi'\sigma'\sigma^{-1}(j) = \pi(j) .$$

Using this condition and Lemma 2.19 once again, a small calculation shows that the equation of H expressed in the $y^{\pi'}$ coordinates is:

$$\text{(i)} \qquad y_{n+1}^{\pi'} - y_{\sigma'(\sigma^{-1}(j))}^{\pi'} = \frac{\mu-k}{\mu}$$

$$\text{or} \quad \text{(ii)} \qquad y_{\sigma'(\sigma^{-1}(j))}^{\pi'} = \frac{k}{\mu} .$$

Next we consider the case where σ and τ belong to different simplexes say $\sigma \subset \Delta^{\pi'}{}_{o}$. Obviously $\Delta^{\pi}{}_{o}$ and $\Delta^{\pi'}{}_{o}$ are adjacent simplexes.

In this case the hyperplane H cuts "nicely" through the subdivision whether it is good or bad so we are only left with c), i.e., $H \cap \Delta_Q$ is a <u>bad</u> hyperplane.

If $\{P_1^{\pi_o}, \cdots, P_{n+1}^{\pi_o}, Q\}$ and $\{P_1^{\pi'_o}, \cdots, P_{n+1}^{\pi'_o}, Q\}$ are the vertexes of Δ^{π_o} and $\Delta^{\pi'_o}$ respectively it follows since H is bad that

$$P_i^{\pi_o} = P_i^{\pi'_o} \qquad \text{for all } i \neq k$$

where k is some integer $2 \leq k \leq n$, Lemma 2.11.

Since σ has a face in common with Δ^{π_o} and τ has a face in common with $\Delta^{\pi'_o}$. This face has to be the common face of $\Delta^{\pi'_o}$ and Δ^{π_o}, namely the face opposite $P_k^{\pi_o}$ (resp. $P_k^{\pi'_o}$). Now σ is determined by a map, say

$$\pi: \{0,1,\cdots,n+1,n+2\} \longrightarrow \{0,1,\cdots,\mu-1\}$$

when $\pi(0) = 0$, $\pi(n+2) = \mu-1$ and τ is determined by <u>the same map</u>, i.e.,

$$\sigma = (\Delta^{\pi_o})^{\pi} \quad \text{and} \quad \tau = (\Delta^{\pi'_o})^{\pi} .$$

Let ρ be the permutation associated to π. Then in the y^{π_o} and $y^{\pi'_o}$ coordinates the vertexes of σ and τ are given by

$$y_k^{\pi_o}(P_j^{\sigma}) = \frac{\pi\rho(k)}{\mu} + \delta_j(k)$$

$$y_k^{\pi'_o}(P_j^{\tau}) = \frac{\pi\rho(k)}{\mu} + \delta_j(k)$$

Suppose that P_j^σ is the vertex of σ that does not meet τ. The equation of the face opposite P_j^σ is then given by Lemma 2.10

$$y^{\pi\sigma}_{\rho^{-1}(j)} - y^{\pi\sigma}_{\rho^{-1}(j-1)} = \frac{1}{\mu}(\pi(j) - \pi(j-1)) = 0 .$$

Hence $\rho^{-1}(j) = k$, $\rho^{-1}(j-1) = k-1 \geq 1$ and

$$\pi(j) = \pi(j-1) .$$

Since $\rho^{-1}(j-1) \geq 1$ it follows that $j \geq 2$. Since $\rho^{-1}(j) = k \leq n$ it follows that $j \not\geq n+2$, i.e.,

$$2 \leq j \leq n+1 .$$

Using the formula 2.19 we get:

$$y_i(P_i^\sigma) = \frac{1}{\nu}\left[\pi_o\sigma_o(j)(1-\ell) + \left(\frac{\pi\rho\sigma_o(j)}{\mu} + \delta_i(\sigma_o(j))\right)\right]$$

$$y_j(P_i^\tau) = \frac{1}{\nu}\left[\pi_o'\sigma_o'(j)(1-\ell) + \left(\frac{\pi\rho\sigma_o'(j)}{\mu} + \delta_i(\sigma_o'(j))\right)\right] .$$

Since $\pi_o'\sigma_o' = \pi_o\sigma_o$ we have

$$y_j(P_i^\sigma) = \varphi(j) + \frac{1}{\nu}\left(\frac{\pi\rho\sigma_o(j)}{\mu} + \delta_i(\sigma_o(j))\right)$$

$$y_j(P_i^\tau) = \varphi(j) + \frac{1}{\nu}\left(\frac{\pi\rho\sigma_o'(j)}{\mu} + \delta_i(\sigma_o'(j))\right)$$

We want to show that the four points $P^{\sigma}_{j-1}\ P^{\sigma}_{j}\ P^{\tau}_{j}\ P^{\sigma}_{j+1}$ form a quadrilateral :

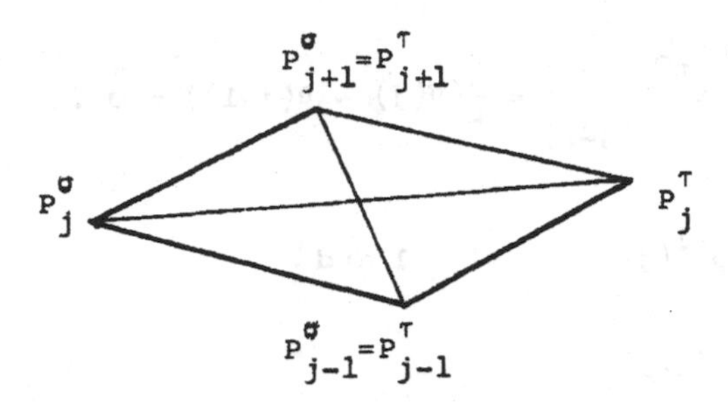

From these formulas it follows that

$$y_s(P^{\sigma}_{j-1}) = y_s(P^{\sigma}_{j}) = y_s(P^{\tau}_{j}) = y_s(P^{\sigma}_{j+1})$$

for all $s \notin \{\sigma_o^{-1}(k), \sigma_o^{-1}(k-1)\}$.

<u>Case $s = \sigma_o^{-1}(k)$.</u>

In this case we have

$\underline{\sigma_o(s) = k}$. $\quad \underline{\sigma_o'(s)} = \epsilon\sigma_o(s) = \epsilon(k) = \underline{k-1}$

$\underline{\rho\sigma_o(s) = j}$ and $\underline{\rho\sigma_o'(s) = j-1}$.

Since $\pi(j) = \pi(j-1)$ we have

$$y_s(P^{\sigma}_{j-1}) = \beta + \frac{1}{\nu}\delta_{j-1}(\sigma_o(s)) = \beta + \frac{1}{\mu\nu}$$

$$y_s(P^{\sigma}_{j+1}) = \beta + \frac{1}{\nu}\delta_{j+1}(\sigma_o(s)) = \beta$$

$$y_s(P^{\sigma}_{j}) = \beta + \frac{1}{\nu}\,\delta_j(\sigma_o(s)) = \beta + \frac{1}{\mu\nu}$$

$$y_s(P^{\tau}_{j}) = \beta + \frac{1}{\nu}\,\delta_j(\sigma'_o(s)) = \beta \quad ,$$

for some constant β.

Hence $y_s(P^{\sigma}_{j-1}) + y_s(P^{\sigma}_{j+1}) = y_s(P^{\sigma}_{j}) + y_s(P^{\tau}_{j})$.

The case $s = \sigma_o^{-1}(k-1)$ we leave to the reader.

This completes the proof of c) since $n+1 \notin \{\sigma_o^{-1}(k), \sigma_o^{-1}(k-1)\}$.

q.e.d.

I believe that we have a quadrilateral also in the case where σ and τ lie in two adjacent simplexes Δ^{π} and $\Delta^{\pi'}$ of type $k = 1$; computations however are more complicated.

Observation 2.21. Let Δ be as in Lemma 2.20 and let H be the hyperplane defined by $\ell = k/\mu$. Denote by P'_i the intersection of H with the line segment P_iQ.

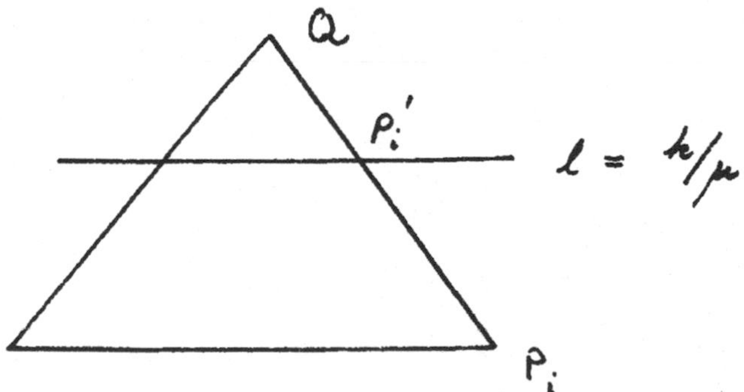

<u>The mixed</u> (ν,μ)<u>-subdivision of Δ induces the mixed</u> $(\nu,\mu-k)$<u>-subdivision of $\Delta' = \{P_1',\cdots,P_{n+1}',Q\}$, and it induces the</u> $\nu\cdot(\mu-k)$<u>-regular subdivision on $\Delta_Q' = \{P_1',\cdots,P_{n+1}'\}$.</u>

<u>Proof</u>. Observation 2.17 and Cor. 3.9.

§2D

<u>Final Example</u>: <u>Global mixed</u> (ν_i,μ_i)<u>-subdivisions</u>

We are now ready to prove the main result of this section. We consider a <u>simplicial</u> complex X with a totally ordered set of vertices $P_1,P_2,\cdots,P_N,Q_{N+1},\cdots,Q_{N+s}$ and such that the open stars of $Q_{N+1},\cdots,Q_{N+s}$ are disjoint. Let $\nu_1,\cdots,\nu_s;\ \mu_1,\cdots,\mu_s$ be a set of positive integers such that

$$\nu_i\mu_i = \nu_j\mu_j = \mu \quad \text{for all } i,j.$$

Let X' be the subdivision of X obtained as follows:

If Δ is a simplex in the open star of Q_{N+i} we let Δ' be the mixed (ν_i,μ_i) subdivision of Δ with respect to the given ordering.

If Δ does not contain any of the Q_i's we let Δ' be the μ-regular subdivision of Δ with respect to the given ordering.

Example:

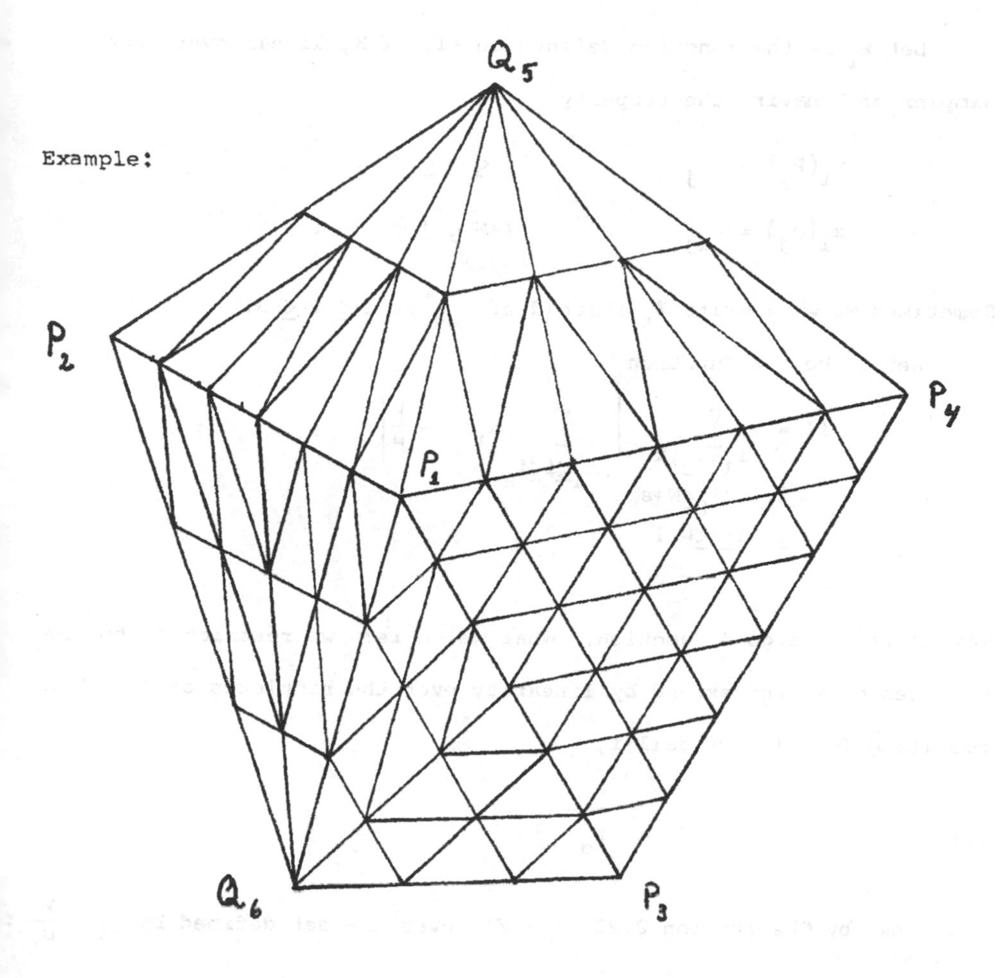

$\nu_1 = 2, \mu_1 = 3, \quad \nu_2 = 3, \mu_2 = 2, \quad \mu = 6$

These subdivisions clearly patch together. Moreover:

<u>Theorem</u> 2.22. <u>X' is a projective subdivision of X.</u>

<u>Proof</u>. The idea of the proof is simply to write down a function and prove that it is good using the numerical criterion.

Let x_i be the function defined on all of X, linear over each simplex of X having the property

$$x_i(P_j) = \delta_{ij} \qquad 1 \leq j \leq N$$

$$x_i(Q_j) = \delta_{ij} \qquad 1+N \leq j \leq N+s .$$

Sometimes we will write ℓ_i instead of x_{N+i}, $1 \leq i \leq s$.

Let f' be the function

$$f' = -\sum_{\substack{i_1,i_2,k \\ 1\leq i_1<i_2\leq N+s \\ 1\leq k\leq \mu-1}} \left| \sum_{i_1\leq i\leq i_2} x_i - \frac{k}{\mu}\right|$$

Now f' is not a good function. What we do is: we restrict f' to the vertices of X' and extend by linearity over the simplexes of X'. The resulting function we call f,

i.e., $$f = \overline{f'_o}$$

Note that by Observation 2.21 $f = f'$ over the set defined by $\ell_i = \frac{k}{\mu_i}$:

i.e.

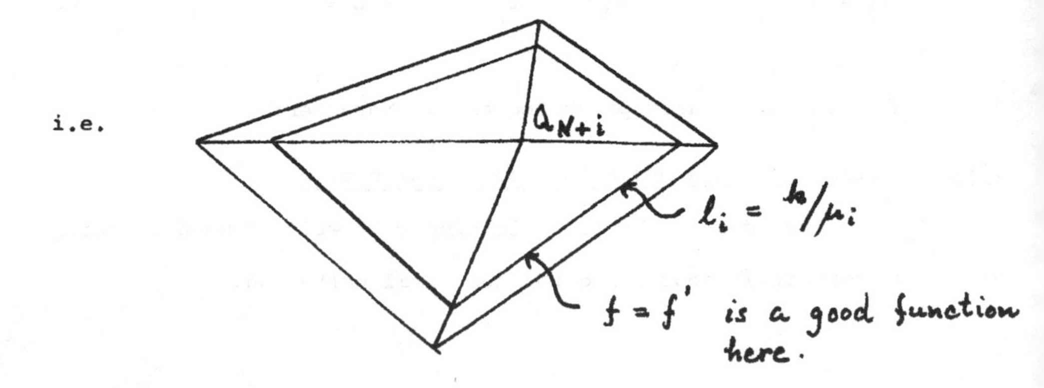

For all pairs of integers m,k such that $1 \leq m \leq N$, $1 \leq k \leq \mu-1$, choose α_i and β_i, $1 \leq i \leq s$, such that

$$\frac{\alpha_i}{\nu_i} \leq \frac{k}{\mu} \leq \frac{\alpha_i+1}{\nu_i} \quad \text{and} \quad \frac{\alpha_i}{\nu_i} + \frac{\beta_i}{\mu} = \frac{k}{\mu} .$$

Define

$$g^1_{m,k} = \begin{cases} -\left| \sum_{j=1}^{m} x_j - \frac{k}{\mu} \right| & \text{if } \ell_i = 0, \quad 1 \leq i \leq s \\ -\left| \sum_{j=1}^{m} x_j + \ell_i \frac{\alpha_i+1}{\nu_i} - \frac{k}{\mu} \right| & \text{if } 0 \leq \ell_i \leq \frac{\beta_i}{\mu_i}, \quad \ell_j = 0 \quad j \neq i \\ -\left| \sum_{j=1}^{m} x_j + \ell_i \frac{\alpha_i}{\nu_i} - \frac{\alpha_i}{\nu_i} \right| & \text{if } \frac{\beta_i}{\mu_i} \leq \ell_i \leq 1, \quad \ell_j = 0 \quad j \neq i \end{cases}$$

and

$$g^2_{m,k} = \begin{cases} -\left| \sum_{j=1}^{m} x_j - \frac{k}{\mu} \right| & \text{if } \ell_i = 0, \quad 1 \leq i \leq s \\ -\left| \sum_{j=1}^{m} x_j + \ell_i \frac{\alpha_i}{\nu_i} - \frac{k}{\mu} \right| & \text{if } 0 \leq \ell_i \leq 1 - \frac{\beta_i}{\mu_i}, \quad \ell_j = 0 \quad j \neq i \\ -\left| \sum_{j=1}^{m} x_j + \ell_i \frac{\alpha_i+1}{\nu_i} - \frac{\alpha_i+1}{\nu_i} \right| & \text{if } 1 - \frac{\beta_i}{\mu_i} \leq \ell_i \leq 1, \quad \ell_j = 0 \quad j \neq i \end{cases}$$

Then put

$$g = \sum_{\substack{0 \leq m \leq N \\ 1 \leq k \leq \mu-1 \\ \alpha=1,2}} g^{\alpha}_{m,k}$$

Comparing this with the equations on page 137 we note that g "breaks" exactly along all the good hyperplanes of X, and possibly along hyperplanes of the form ℓ_i = constant.

Finally we define h by

$$h = -\sum_{i=1}^{s} \sum_{1 \leq k \leq \mu_i - 1} \left| \ell_i - \frac{k}{\mu_i} \right|$$

If ϵ_1 and ϵ_2 are positive real numbers put

$$F_{\epsilon_1 \epsilon_2} = h + \epsilon_1 g + \epsilon_2 f \quad .$$

For each pair of adjacent (max dimensional) simplexes σ and τ of X' such that σ and τ belong to the same simplex of X, we choose a little line segment through the common face and define

$$\Delta_{\sigma\tau}(f)$$

as in the proof of Lemma 1.8.

There are 4 mutually exclusive ways of choosing such pairs, and these 4 cases exhaust all possibilities.

<u>Case 1</u>. σ and τ belong to a simplex of X and this simplex does not contain Q_i as a vertex $N+1 \leq i \leq N+s$. We clearly have

$$\Delta_{\sigma\tau}(f) \geq \lambda_2 > 0 \qquad \text{for some constant } \lambda_2$$

$$\Delta_{\sigma\tau}(g) \geq 0$$

$$\Delta_{\sigma\tau}(h) \geq 0.$$

Hence

$$\Delta_{\sigma\tau}(F_{\epsilon_1\epsilon_2}) \geq \epsilon_2\lambda_2 \quad .$$

<u>Case 2</u>. σ and τ lie in a simplex which has Q_i as a vertex, and the hyperplane that separates σ and τ is of the form ℓ_i = constant

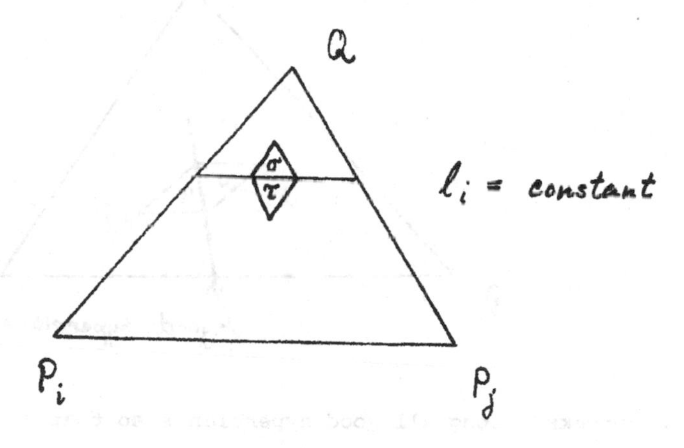

Clearly

$$|\Delta_{\sigma\tau}(f)| < K_2 \quad \text{for some constant } K_2$$

$$|\Delta_{\sigma\tau}(g)| < K_1 \quad \text{for some constant } K_1$$

$$\Delta_{\sigma\tau}(h) \geq \lambda_o > 0 \quad \text{for some constant } \lambda_o.$$

Hence we have

$$\Delta_{\sigma\tau}(F_{\epsilon_1\epsilon_2}) \geq \lambda_o - \epsilon_1 K_1 - \epsilon_2 K_2$$

<u>Case 3.</u> σ and τ lie in a simplex which has Q_i as a vertex, and moreover the hyperplane that separates σ and τ intersects the bottom face in a <u>good</u> hyperplane,

i.e.,

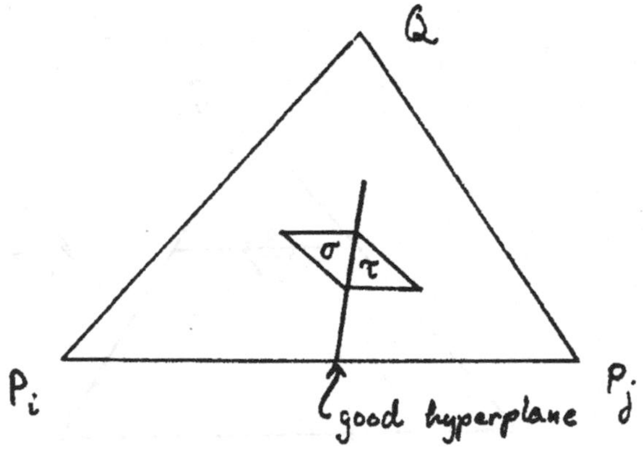

Now g "breaks" along all good hyperplanes so that

$$|\Delta_{\sigma\tau}(f)| < K_2$$

$$\Delta_{\sigma\tau}(g) \geq \lambda_1 > 0 \qquad \text{for some constant } \lambda_1$$

$$\Delta_{\sigma\tau}(h) = 0$$

and we have

$$\Delta_{\sigma\tau}(F_{\epsilon_1\epsilon_2}) \geq \epsilon_1\lambda_1 - \epsilon_2 K_2$$

<u>Case 4</u>. Again σ and τ lie in a simplex of X which has Q_i as a vertex but this time the hyperplane that separates σ and τ intersects the bottom face in a bad hyperplane. We have

$\Delta_{\sigma\tau}(f) \geq \lambda_2 > 0$ by Lemma 2.20, a)

$\Delta_{\sigma\tau}(g) = 0$ this because g is linear except across good hyperplanes or ℓ_i = const. hyperplanes.

$\Delta_{\sigma\tau}(h) = 0.$

Hence

$$\Delta_{\sigma\tau}(F_{\epsilon_1\epsilon_2}) \geq \epsilon_2\lambda_2$$

By choosing ϵ_1 and ϵ_2 such that $0 < \epsilon_1 K_1 < \lambda_o/2$ and $0 < \epsilon_2 K_2 < \min\{\lambda_o/2, \epsilon_1\lambda_1\}$ which is clearly possible, we see that

$$\Delta_{\sigma,\tau}(F_{\epsilon_1\epsilon_2}) > 0$$

in all the cases. The theorem now follows from Lemma 1.8.

q.e.d.

§2E

We conclude this section by giving two rather simple examples of non-projective subdivisions.

Example 1 (Hironaka).

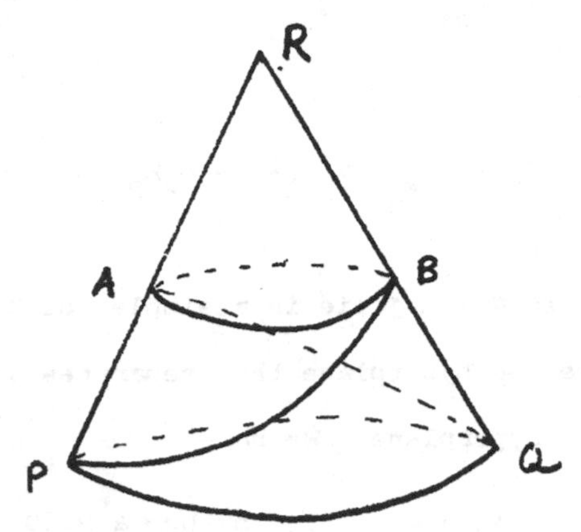

This is two triangles which are glued together along the edges PR and QR (Hence as a complex not embeddable in $\mathbb{R}^n$). Suppose f is a good function for this subdivision. Adding to f a function linear on each triangle we may suppose that $f(P) = f(Q) = f(R) = 0$. If f is good on the front triangle we necessarily get $f(B) > f(A)$. But looking at the back we must have $f(A) > f(B)$. Hence the subdivision is non-projective. This example corresponds to the following "blow-up":

PR and QR "correspond" to two lines ℓ and m in $\mathbb{P}^3$

intersecting in two points S and T "corresponding" to the faces PQR fro and PQR back resp.

Now the subdivision corresponds to blowing up the two lines outside S and T. However at S we first blow up the line m and then blow up the proper transform of ℓ and at T we reverse the order. Of course we can glue these things together algebraically. The result yields a non-singular, non-projective complete algebraic threefold first discovered by Hironaka (Annals of Math., 75 (1962), p.190).

Example 2. (F. Commoner)

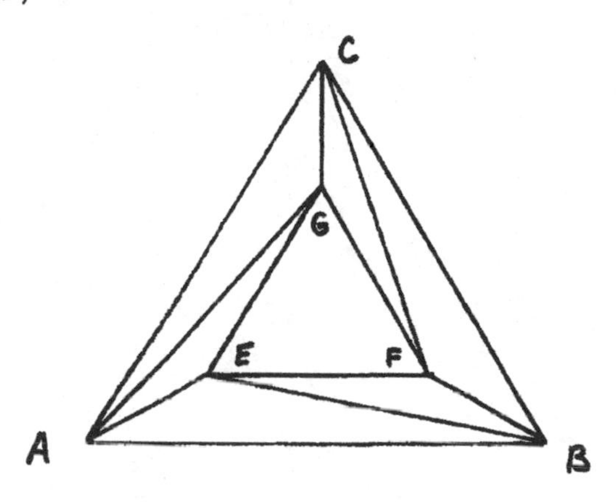

Again if f is a good function, we can assume $f(A) = f(B) = f(C) = 0$; but then we must have

$$f(E) > f(F) > f(G) > f(E) .$$

The geometric analogue in this case yields a proper non-projective birational morphism

$$X \xrightarrow{\ \pi\ } \mathbb{A}^3$$

such that π is an isomorphism on $X - \pi^{-1}(0)$.

§3. Waterman points

In this section we fix the following:

Δ is a simplex in $\mathbb{R}^n$ with an ordered set of integral vertices $P_o, \ldots, P_n$ and we denote the vector $\overrightarrow{P_o P_i}$ by e_i. The e_i's generate a lattice which we call L.

Considering Δ as a simplicial complex we get a rational structure on Δ by restricting all integral combinations of 1 and the coordinate functions to Δ. We note this rational structure by M. Moreover we suppose that the multiplicity of Δ with respect to M and the integer 1 is $k > 1$, i.e.,

$$m(\Delta, M, 1) = k .$$

The quotient group $\mathbb{Z}^n/L$ we denote by $W(\Delta)$. Clearly we have

$$\#(W(\Delta)) = k.$$

If M_o denotes the set of linear plus constant functions which take integral values on the vertices of Δ we have $M \subset M_o$ and the evaluation map gives us a pairing

$$M_o/M \times W(\Delta) \longrightarrow \mathbb{Q}/\mathbb{Z}$$

$$f(e_i) \longmapsto f(P_i) - f(P_o).$$

By definition $M_o/M \simeq \widehat{W(\Delta)}$ and therefore $\#(M_o/M) = k$ also.

If $[\omega]$ is a nonzero element of $W(\Delta)$, $[\omega]$ has a unique representation as a vector

$$\omega = \sum_{i=1}^{n} \alpha_i e_i \qquad \text{where } 0 \leq \alpha_i < 1 .$$

We define $\nu([\omega]) = \left\{ \sum \alpha_i \right\}$ where

$$\{\beta\} = \min\{n \in \mathbb{Z},\ n \geq \beta\},$$

i.e., $\{\beta\} = -[-\beta]$.

The point $P([\omega]) = \omega/\nu([\omega]) + P_o$ clearly lies in Δ so P is a mapping

$$P\colon W(\Delta) - (0) \longrightarrow \Delta,$$

but not necessarily 1-1.

<u>Definition 3.1. The points in the image of P will be called the waterman points of Δ.</u>

<u>For any waterman point</u> $P([\omega])$ <u>in Δ, let</u> $[\omega],[\omega_1],[\omega_2],\cdots,[\omega_s]$ <u>be the full inverse image of $P([\omega])$. The integer</u>

$$\nu = \min\{\nu([\omega]),\nu([\omega_1]),\cdots,\nu([\omega_s])\}$$

<u>will be called the value of the waterman point.</u>

Note that for a ν-valued waterman point $P = P([\omega])$, $\nu \cdot P$ is an integral point.

Let P be a ν-valued waterman point in Δ. For each $i \in \{0,\cdots,n\}$ such that P does not belong to the face $\{P_0,P_1,\cdots,\hat{P}_i,\cdots,P_n\}$ we subdivide the simplex $\{P_0,\cdots\hat{P}_i\cdots P_n\}$ regularly with respect to this ordering and the integer ν, and take the cones over the simplices of this subdivision with apex P. This gives us a rational subdivision Δ' of Δ. The following lemma is due to Alan Waterman:

Lemma 3.2. With the above notations we have (Δ',M) integral over $\frac{1}{\nu}\mathbb{Z}$ and

$$m(\Delta',M,\nu) < k.$$

Proof. We have $\nu\cdot(P-P_o) = \alpha_1 e_1+\alpha_2 e_2+\cdots+\alpha_n e_n$ where for all i, $0 \leq \alpha_i < 1$.

If Γ is a simplex of Δ and a face of Γ lies in the i-th face of Δ for $1 \leq i \leq n$ we have

$$\mathrm{vol}(\Gamma) = \frac{1}{\nu^n\cdot n!}\left|\det(e_1\cdots,\alpha_i e_i,\cdots e_n)\right|$$

and hence

$$m(\Gamma,M,\nu) = \alpha_i\cdot k < k.$$

If a face of Γ lies in the face opposite P_o we get

$$m(\Gamma,M,\nu) = (\nu - \sum\alpha_i)\cdot k$$

and $\nu - \sum\alpha_i < 1$ by the very definition of ν. q.e.d.

We can illustrate this lemma by a simple example: let Δ be the simplex in $\mathbb{R}^3$ given by the coordinates $\{(0,0,0)(100)(010)(1,1,3)\}$. Clearly

$$m(\Delta,M,1) = 3$$

and we have two 2-valued waterman points $P = (\frac{1}{2}\,\frac{1}{2}\,\frac{1}{2})$, $Q = \{\frac{1}{2},\frac{1}{2},1\}$

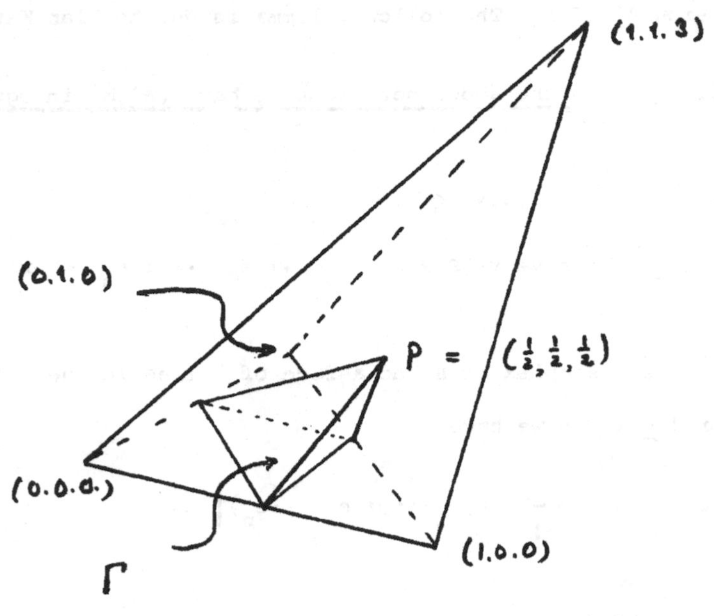

If Δ' is the above subdivision of Δ with respect to P we have

$$m(\Delta',M,2) = 2.$$

In particular if Γ is as in the picture above we have

$$m(\Gamma,M,2) = 1 .$$

<u>Remark</u>. <u>Note that we can always find a waterman point P of value $\nu \leq \{n/2\}$.</u>

Lemma 3.3. Let Δ_α be any face of Δ, then Δ_α and Δ have the same multiplicity if and only if all the waterman points of Δ lie in the face Δ_α.

Proof. It is clear that for any α we have injections

$$W(\Delta_\alpha) \longrightarrow W(\Delta).$$

To say that all the waterman points actually lie in Δ_α means that the above injection is an isomorphism and so the two groups have the same cardinality, i.e., Δ_α and Δ have some multiplicity. The converse is also clear.

Corollary 3.4. Let Δ' and Δ'' be two faces of Δ such that $m(\Delta') = m(\Delta'') = m(\Delta) = k$ then $m(\Delta' \cap \Delta'') = k$.

This last corollary gives us a remarkable decomposition theorem.

Proposition 3.5. Let X be a simplicial complex with a rational structure L and μ an integer such that the functions in L take values in $1/\mu\mathbb{Z}$ on the vertices of X.

Suppose $m(X,L,\mu) = k > 1$ and let

$$U = \bigcup_{m(\sigma_\alpha,L,\mu)=k} \operatorname{int}(\sigma_\alpha) .$$

Then U breaks up into connected components U_i each being the open star neighborhood of a simplex of multiplicity k, i.e.,

$$U = \bigcup_{\text{disjoint}} \operatorname{star}(\sigma_i).$$

We end this section with a discussion of the existence of waterman points in the interior of a simplex.

Lemma 3.6. *Let Δ be as before and suppose that*

i) *$W(\Delta)$ is a cyclic group.*

ii) *For all proper faces Δ_α of Δ* $W(\Delta_\alpha) < k$.

Then if $[\omega]$ is a generator of $W(\Delta)$

$$P([\omega]) \in \text{int } \Delta.$$

Proof. If $P([\omega])$ belonged to some face, say Δ_α, we would have $[\omega] \in W(\Delta_\alpha)$.

Since $[\omega]$ is a generator of $W(\Delta)$ and $W(\Delta_\alpha)$ injects into $W(\Delta)$ this would imply that $W(\Delta_\alpha) \cong W(\Delta)$ contradicting ii). q.e.d.

Corollary 3.7. *Let X be as in Proposition 3.5 and let $\sigma_1 \cdots \sigma_s$ be the simplexes such that*

$$U = \bigcup_{\text{disjoint}} \text{star}(\sigma_i) .$$

Then if k is a prime number each of the simplices σ_i have interior waterman points.

§4. Statement and proof of the main theorem

Theorem 4.1 (Main Theorem) Let X be a polyhedral complex, L a rational structure, and μ an integer such that (X,L) is integral over $\frac{1}{\mu}\mathbb{Z}$.

Then there exists an integer ν and a rational projective subdivision X' of X such that (X',L) is integral over $\frac{1}{\mu\cdot\nu}\mathbb{Z}$ and

$$m(X',L,\mu\cdot\nu) = 1 \quad .$$

Proof. By the transitivity of projective subdivisions we may as well suppose that X is simplicial since the barycentric subdivision is rational and projective. We will prove the theorem with induction on the number $k = m(X,L,\mu)$. So suppose the theorem is true for all simplicial complexes X'' such that $m(X'',L'',\mu'') < k$. For the inductive step we divide into two cases.

Case 1. k is a composite number.

For each simplex σ_i of multiplicity k we pick out a waterman point $P_i \in \sigma_i$, say of value ν_i and of order $p_i < k$, p_i a prime number and $1 \leq i \leq s$. Then we define a decreasing sequence of rational structures $M_o \supset M_1 \supset \cdots \supset M_s \supset L$ as follows:

$$M_o = \left\{ \begin{array}{l} \text{all linear functions which take values} \\ \text{in } \frac{1}{\mu}\mathbb{Z} \text{ on the vertices of X} \end{array} \right\}$$

$$M_1 = \left\{ \begin{array}{l} \text{all linear functions } f \in M_o \text{ with the} \\ \text{extra condition } \quad f(P_1) \in \frac{1}{\mu\cdot\nu_1}\mathbb{Z} \end{array} \right\}$$

$$M_i = \begin{Bmatrix} \text{all linear functions } f \in M_{i-1} \text{ with the} \\ \text{extra condition } f(P_i) \in \frac{1}{\mu \cdot \nu_i} \mathbb{Z} \end{Bmatrix}$$

Since M_i is obtained from M_{i-1} by one extra condition it follows by duality that for any subdivision X" of X and any simplex $\sigma_\alpha \in |X"|$ we have

$$\# (M_{i-1,\alpha}/M_{i,\alpha}) = p_i \quad \text{or} \quad 1 \ .$$

Note also that for each i

$$p_i \,|\, m(\sigma_i, M_i, \mu)$$

hence

$$m(\sigma_i, M_s, \mu) > 1 \ .$$

Now if X" is a projective subdivision of X such that $(X", M_{i-1})$ is integral over $\frac{1}{\mu\nu"}\mathbb{Z}$ and

$$m(X", M_{i-1}, \mu\nu") = 1$$

we will have $(X", M_i)$ integral over $\frac{1}{\mu\nu"}\mathbb{Z}$ and

$$m(X", M_i, \mu\nu") \leq p_i < k \ .$$

So by induction we can find a projective subdivision X''' of X" and an integer ν''' such that (x''', M_i) is integral over $\frac{1}{\mu}\nu"\nu'''\mathbb{Z}$ and

$$m(X''',M_i\cdot\mu\nu''\nu''') = 1 .$$

By induction and transitivity of projective subdivisions we may proceed with this process all the way to s, i.e., there is a projective subdivision X" of X and a number ν'' such that (X'',M_s) is integral over $\frac{1}{\mu}\nu''\mathbb{Z}$ and

$$m(X'',M_s,\mu\nu'') = 1 .$$

Since $M_s \supset L$, (X',L) is integral over $\mu\nu''$ as well and moreover all simplices of multiplicity k with respect to L and μ have been decomposed into simplices with lower multiplicity with respect to L and $\mu\nu''$, cf. Observation 1.2.

Hence

$$m(X'',L,\mu\nu'') < k$$

and the theorem follows by induction.

<u>Case</u> 2. <u>k = p = prime number</u>.

Let U be the union of all the interiors of simplices of multiplicity p. Then as in Corollary 3.7, U splits up into a disjoint union:

$$U = \bigcup_{\text{disjoint}} \text{star}(\sigma_i) .$$

And each σ_i has an interior waterman point Q_i, say of value ν_i. Let μ_i be integers such that $\nu_1\mu_1 = \nu_2\mu_2 = \nu_3\mu_3 = \cdots = \nu_i\mu_i = \cdots = \nu$, $1 \leq i \leq s$, and let X' be the mixed (ν_i, μ_i) subdivision of X with respect to the points Q_i, $1 \leq i \leq s$.

Then X' is projective and all the simplices of multiplicity p have been refined, hence

$$m(X', L, \mu \cdot \nu) < p,$$

and the theorem follows by induction. q.e.d.

Chapter IV

Further Applications

§1. Actions of T^{n-1} on X^n

The purpose of this section is to study n-dimensional normal varieties X^n on which an (n-1)-dimensional torus T^{n-1} is acting faithfully. Surprisingly, we can describe this situation almost as well as when the torus is n-dimensional. But if the torus were only (n-2)-dimensional, we wouldn't be able to do anything! When $n = 2$, this situation has been studied by Orlik and Wagreich (Annals of Math., 93 (1971), p. 205; Math. Annalen, 193 (1971), p. 121) and our results generalize some, but by no means all of theirs to higher dimensions. They have quite a different generalization to higher dimensions that we will mention later.

Step I: We can apply standard results on the existence of birational orbit spaces (see for instance SGA 3, exposé V, Th. 8.1) to show that there is a non-empty open set $W \subset X$ stable under T and a smooth morphism:

$$\pi: W \longrightarrow U$$

(U non-singular curve) making U into an orbit space W/T: even more, so that

$$W \times_U W \cong T \times W.$$

If $\eta \in U$ is the generic point, it follows that $\pi^{-1}(\eta)$ is a principal homogeneous space over the torus $T \underset{\text{Spec } k}{\times} \text{Spec } k(\eta)$: since all such p.h. spaces split by Hilbert's Theorem 90 (see for instance, Springer's article on Galois Cohomology in <u>Algebraic groups and Discontinuous subgroups</u>, AMS Symposia IX), it follows that $\pi^{-1}(\eta) \cong T \underset{\text{Spec } k}{\times} \text{Spec } k(\eta)$, hence replacing U by a smaller open set if necessary, $W \cong T \times U$. In other words, we have an embedding situation:

$$T \times U \subset X$$

open
T-equivariant
embedding

(T acting trivially on U).

<u>Step II</u>: Let C be the complete non-singular curve in which U is an open set. For all $x \in C$, let

$M(x)$ = value group $k(C)^*/\mathcal{O}^*_{x,C}$, or the group of divisors $n\cdot(x)$, $n \in \mathbb{Z}$

$N(x) = \mathrm{Hom}(M(x), \mathbb{Z})$

$N_{\mathbb{R}}(x) = \mathrm{Hom}(M(x), \mathbb{R})$

$N^+_{\mathbb{R}}(x) \subset N_{\mathbb{R}}(x)$, the maps taking positive divisors to $\mathbb{R}^+$.

Assume for this step that the rational map $X \cdots\!> C$ extending $p_2: T \times U \longrightarrow U$ is a morphism:

$$\pi: X \longrightarrow C .$$

In this case, I claim:

<u>Theorem</u>. <u>After possibly shrinking</u> U, $T \times U \subset X$ <u>is a toroidal embedding without self-intersection. Moreover, its graph</u> Δ <u>can be canonically embedded as follows:</u>

$$\Delta \subset \bigcup_{x \in C-U} N_{\mathbb{R}}(T) \times N_{\mathbb{R}}^{+}(x) \Big/ \left(\begin{matrix}\text{glued together} \\ \text{along } N_{\mathbb{R}}(T) \times (0)\end{matrix}\right)$$

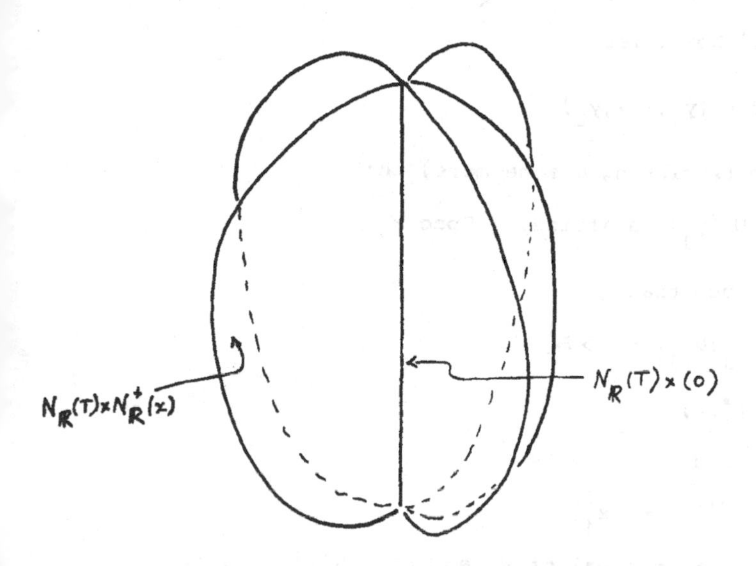

<u>Proof</u>: By Sumihiro's Theorem, we can cover X by T-invariant affine open sets $X_{(i)}$. Then $X_{(i)} \cap (T \times U)$ is T-invariant, hence of the form $T \times U_i$ for some $U_i \subset U$. Replacing U by $\bigcap U_i$, we may assume $X_{(i)} \supset T \times U$. Now if

$$X_{(i)} = \text{Spec } S_i, \quad U = \text{Spec } R$$

then

$$S_i \subset \Gamma(T \times U, \mathcal{O}_X) = \bigoplus_{\alpha \in M(T)} R \cdot \mathfrak{X}^{\alpha} .$$

Since $X_{(i)}$ is T-invariant, the subring S_i is also graded, i.e.,

$$S_i = \bigoplus_{\alpha \in M(T)} T_\alpha \cdot \mathfrak{X}^\alpha, \qquad T_\alpha \subset R.$$

Therefore S_i is generated over k by a finite set of monomials $c_k \cdot \mathfrak{X}^{\alpha_k}$, $c_k \in R$. Now we may shrink U if necessary even further, replacing R by localizations, until $c_k \in R^*$ for all generators $c_k \cdot \mathfrak{X}^{\alpha_k}$ of all S_i. Next, let

$$C-U = \{y_1, \cdots, y_t\}.$$

We may also assume (shrinking U some more) that

a) $U_{(j)} = U \cup \{y_j\}$ is affine, $=$ Spec R_j

b) $\exists\, \pi_j \in R_j$ such that:

$$\rho_j : U_{(j)} \longrightarrow \mathbb{A}^1_k$$

$$\rho_j^*(t) = \pi_j$$

is étale and

$$\rho_j^{-1}(0) = \{x_j\}$$

(hence $\pi_j \cdot R_j =$ ideal of y_j and $R^* = R_j^* \times \{\pi_j^k\}_{k \in \mathbb{Z}}$).

Finally X is covered by the open sets $X_{(i)} \cap \pi^{-1}(U_{(j)})$ which are affine with rings which may be presented as follows:

$$\Gamma\left(X_{(i)} \cap \pi^{-1}(U_{(j)}), \mathcal{O}_X\right) = R_j\left[\cdots, \pi^{r_\ell} \cdot \mathfrak{X}^{\alpha_\ell}, \cdots\right]$$

for some finite set of pairs $S_{i,j} = \left\{(r_\ell, \alpha_\ell) \in \mathbb{Z} \times M(T)\right\}$. But this implies that $T \times U \subset X$ is a toroidal embedding without self-intersection. To see this, define

$$Y_{(i,j)} = \operatorname{Spec} k[t,\cdots,t^{r_\ell}\cdot \mathfrak{X}^{\alpha_\ell},\cdots]_{(r_\ell,\alpha_\ell)\in S_{ij}}.$$

Then we have an affine torus embedding

$$T\times \mathbb{G}_m \subset Y_{(i,j)}$$

and an isomorphism q:

$$\begin{array}{ccccc} X_{(i)}\cap\pi^{-1}(U_{(j)}) & \xrightarrow{q} & U_{(j)}\times_{\mathbb{A}^1_t} Y_{i,j} & \xrightarrow{p_2} & Y_{i,j} \\ \cup & & \cup & & \cup \\ T\times U & \xrightarrow{\approx} & U_{(j)}\times_{\mathbb{A}^1_t}(T\times\mathbb{G}_m) & \xrightarrow{p_2} & T\times\mathbb{G}_m \end{array}$$

(In particular, since ρ_j is etale, p_2 is etale and this proves that Y_{ij} is normal). Therefore, at all $x\in X_{(i)}\cap\pi^{-1}(U_{(j)})$, there is a formal isomorphism of X near x with $Y_{i,j}$ near $p_2(q(x))$. Also, since every component of $Y_{ij}-T\times\mathbb{G}_m$ is normal, this proves that every component of $X_{(i)}\cap\pi^{-1}(U_{(j)})-T\times U$ is normal.

Next, I claim that if we fix j, but let i vary, the affine pieces $Y_{i,j}$ can be patched together to a big variety Y_j containing $T\times\mathbb{G}_m$ as follows:

$$\begin{array}{ccccc} \pi^{-1}(U_{(j)}) & \xrightarrow{\approx} & U_{(j)}\times_{\mathbb{A}^1_t} Y_j & \longrightarrow & Y_j \\ \| & & \| & & \| \\ \bigcup_i X_{(i)}\cap\pi^{-1}(U_{(j)}) & \xrightarrow{\approx} & \bigcup_i U_{(j)}\times_{\mathbb{A}^1_t} Y_{ij} & \longrightarrow & \bigcup_i Y_{i,j} \\ \cup & & \cup & & \cup \\ T\times U & \xrightarrow{\approx} & U_{(j)}\times_{\mathbb{A}^1_t}(T\times\mathbb{G}_m) & \longrightarrow & T\times\mathbb{G}_m \end{array} .$$

In fact, patchability means that for all i_1, i_2, the isomorphism of $T\times\mathbb{G}_m$ in $Y_{i_1,j}$ with $T\times\mathbb{G}_m$ in $Y_{i_2,j}$ extends to an isomorphism of some $T\times\mathbb{G}_m$-invariant affine $(Y_{i_1,j})_{i_2} \subset Y_{i_1,j}$ with $(Y_{i_2,j})_{i_1} \subset Y_{i_2,j}$ and that this isomorphism has a closed graph in $Y_{i_1,j}\times Y_{i_2,j}$. But we know the closure of the graph of the first isomorphism, when pulled back by $U_{(j)}\times_{\mathbb{A}^1_t}(*)$ to a subset of $\left[X_{(i_1)}\cap\pi^{-1}(U_{(j)})\right]\times\left[X_{(i_2)}\cap\pi^{-1}(U_{(j)})\right]$, is just the graph of the identity map between the common T-invariant affine $X_{(i_1)}\cap X_{(i_2)}\cap\pi^{-1}(U_{(j)})$. This is easily seen to imply patchability.

Thus we have "approximated" X along the T-invariant open set $\pi^{-1}(U_{(j)})$ by an actual torus embedding $T\times\mathbb{G}_m \subset Y_j$. In this "approximation",

a) there is a bijection between strata of $\pi^{-1}(U_{(j)})-T\times U$ and strata of $Y_j - T\times\mathbb{G}_m$,

[In fact, the fibres over $y_j \in U_{(j)}$, $0 \in \mathbb{A}^1_t$ are isomorphic, so these strata correspond; and all strata of Y_j which dominate $\mathbb{A}^1_t$ are of the form $(T/H)\times\mathbb{G}_m$, some subgroup $H \subset T$, and so do not split up when you take fibre product $U_{(j)}\times_{\mathbb{A}^1_t}(*)$.]

b) the divisors $(t^r\cdot\mathfrak{X}^\alpha)$, $(r,\alpha)\in M(\mathbb{G}_m\times T)$, correspond to the divisors $(\pi_j^r\cdot\mathfrak{X}^\alpha)$, $(r,\alpha)\in M(x)\times M(T)$, on $\pi^{-1}(U_{(j)})$.

It follows that the graphs of $(\pi^{-1}(U_{(j)}),T\times U)$ and of $(Y_j,T\times\mathbb{G}_m)$ are isomorphic and that the graph of the former lies canonically in

$N_{\mathbb{R}}(y_j) \times N_{\mathbb{R}}(T)$. In fact since π_j has zeroes and no poles on $\pi^{-1}(U_{(j)})$, this graph lies in $N^+_{\mathbb{R}}(y_j) \times N_{\mathbb{R}}(T)$, in such a way that

c) polyhedra $\sigma \not\subset (0) \times N_{\mathbb{R}}(T)$ correspond to strata $Z \subset \pi^{-1}(y_j)$

d) polyhedra $\sigma \subset (0) \times N_{\mathbb{R}}(T)$ correspond to strata $Z \subset \pi^{-1}(U)$.

Since X is obtained by glueing the open sets $\pi^{-1}(U_{(j)})$ along the common open $\pi^{-1}(U)$, the graph of $(X, T\times U)$ is obtained by glueing the graphs of $(\pi^{-1}(U_{(j)}), T\times U)$ along the common subgraph, namely that of $(\pi^{-1}(U), T\times U)$. In view of (c) and (d), this glueing lies canonically in the union of $N^+_{\mathbb{R}}(y_{(j)}) \times N_{\mathbb{R}}(T)$ over j, identified on $(0) \times N_{\mathbb{R}}(T)$. QED

Step III: We now return to the general case of a T^{n-1} acting on an X^n, but drop the assumption that $X \longrightarrow C$ is a morphism. This is the interesting case: it can be reduced to the simpler case, however, by considering:

$$X' = \text{graph in } X\times C \text{ of the rational map } X \longrightarrow C$$
$$X'' = \text{normalization of } X'.$$

Then T acts on all 3 birational varieties X, X' and X'':

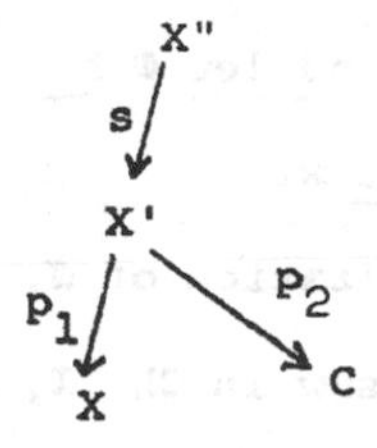

and X''/C satisfies the conditions of Step II. Define

$$S = \left\{ x \in X \,\middle|\, \text{the rational map } X \longrightarrow C \text{ is not defined at } x \right\}$$

$q: X-S \longrightarrow C$ the given morphism.

Then set-theoretically:

$$X' = (X-S) \cup (S \times C)$$

i.e., $p_1^{-1}(x)$ = one pt., if $x \notin S$; $p_1^{-1}(x) \cong C$, if $x \in S$. (This follows because $p_1^{-1}(x)$ is connected and $p_1^{-1}(x) \subset \{x\} \times C$.)
In X", $S \times C$ may be replaced by a finite covering. However we always have:

Proposition: Suppose $x \in X$ is a point fixed by the action of T where $q: X \cdots\to C$ is not defined. Then

$$s^{-1}(p_1^{-1}(x)) \cong C ,$$

i.e., x is blown up into a curve of points isomorphic to C via $p_2 \circ s: X'' \longrightarrow C$. Moreover, this curve is left pointwise fixed by T.

Proof: This depends on the handy lemma:

Lemma: Let a torus T act on $\mathbb{A}^n$ and let $\mathfrak{O}$ be an orbit. If $x \in \overline{\mathfrak{O}}$ is fixed by T, then $\overline{\mathfrak{O}}$ is unibranch at x.

Proof: Let $\overline{\mathfrak{O}}'$ = the normalization of $\overline{\mathfrak{O}}$. Then $\overline{\mathfrak{O}}'$ is an affine normal embedding of T, and as we saw in Ch. I, §1, $\overline{\mathfrak{O}}'$ has exactly one closed T-orbit: hence if x split into several points in $\overline{\mathfrak{O}}'$, each would be a fixed point of T. So x does not split up, and $\overline{\mathfrak{O}}$ is unibranch at x. QED

To prove the theorem, recall that there is an open set $T \times U \subset X$ at which $q: X \cdots > C$ is defined and at which X is non-singular. Hence we find:

$$\begin{array}{ccc} T\times U \subset X'' \\ \| \quad\quad \downarrow \\ T\times U \subset X' \\ \| \quad\quad \downarrow \\ T\times U \subset X \end{array}$$

Moreover, for all but a finite number of $a \in U$, $T\times\{a\}$ will be dense in the fibre over a in both X' and X'':

$$\begin{array}{ccc} T\times\{a\} \subset s^{-1}(p_2^{-1}(a)) = X''_a \\ \| \quad\quad\quad \downarrow \\ T\times\{a\} \subset p_2^{-1}(a) \quad = X'_a \end{array}$$

Therefore X''_a will be finite and birational over X'_a. As $(x,a) \in X'_a$ is left fixed by T, it follows from the lemma that (x,a) has only one point over it in X''_a. As a varies, this shows that the map

$$p_2 \cdot s: \quad s^{-1}(\{x\}\times C) \longrightarrow C$$

is almost everywhere 1-1 as well as everywhere finite-to-one. But by Zariski's connectedness theorem $s^{-1}(\{x\}\times C) = s^{-1}(p_1^{-1}(x))$ is connected too. Thus $s^{-1}(\{x\}\times C) \cong C$ in fact. QED

Now we can combine Steps II and III to describe resolutions for any T-fixed points $x \in X$: by Step II, $T \times U \subset X''$ is a toroidal embedding without self-intersection. Let $C'' \subset X''$ be the curve into which x is blown up under $p_1 \cdot s: X'' \longrightarrow X$. Inside the full graph Δ of $T \times U \subset X''$, the Star of C'' will form a subgraph $\Delta_x \subset \Delta$. The conical polyhedral complex Δ_x with integral structure is an invariant* of the singularity x. To give a rough picture of Δ_x, let $C-U = \{y_1, \cdots, y_t\}$, and let $y_i'' \in C''$ be the point lying over y_i. Then the embedding $T \times U \subset \text{Star}(y_i'')$ defines a polyhedron σ_i, and σ_i has a distinguished face σ_i^o corresponding to the strata Y such that $C'' \subset Y$. Then

$$\Delta_x \cong \bigcup_{i=1}^{t} \sigma_i \Big/ \left(\begin{array}{l} \text{glued together} \\ \text{along their isomorphic} \\ \text{faces } \sigma_i^o \end{array} \right)$$

If we projectivize, i.e., divide this conical complex by $\mathbb{R}_+^*$, we get a compact complex that looks like this for $n = 2$ and 3:

<u>n = 2</u>

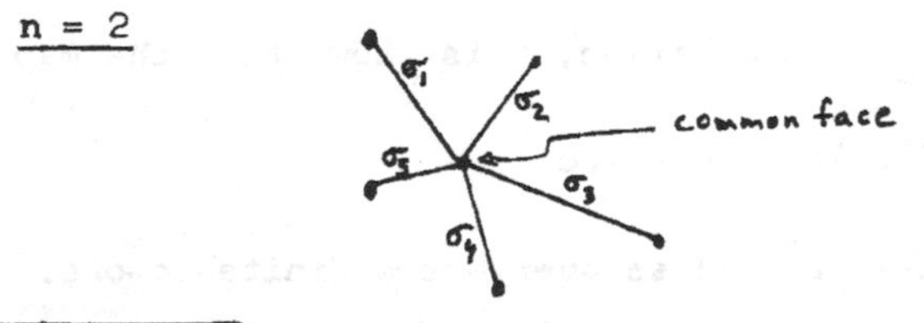

---

*The only choice made was of the open set $U \subset C$: but if U is made smaller, although potentially a new "fin" is added to the complex Δ, one sees this "fin" consists only in the old points in $N_{\mathbb{R}}(T) \times (0)$, hence Δ and Δ_x don't change.

$\underline{n = 3}$

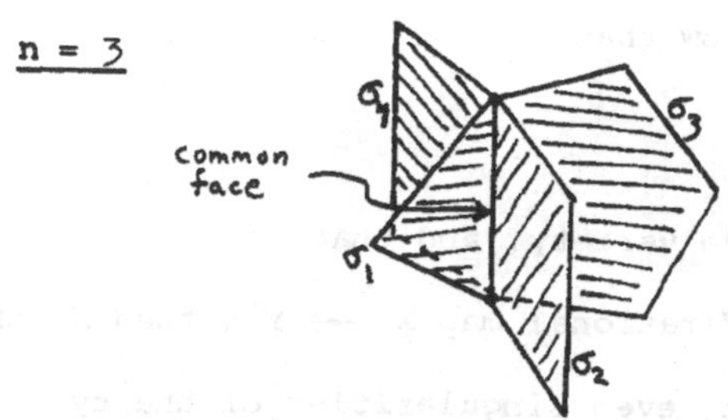

Note that in all cases, the "fins" of Δ_x correspond 1-1 with a finite set of exceptional points $y_i \in C$. By the general theory of Ch. II, §2, all polyhedral subdivisions $\Delta_x = \bigcup \tau_\alpha$, where each τ_α is a simplicial cone generated by a subset of a basis of the integral lattice on the σ_i containing it (see Chapter II, §2, Theorems 4*, 6* and 8*), define resolutions of X near x. Moreover, in some sense, all "canonical" resolutions of X that dominate X" are of this type.

Finally, I want to describe a quite different higher-dimensional generalization of the n = 2 case that Orlik and Wagreich are considering (cf. their article in the Proceedings of the Conference on Transformation Groups, Amherst, 1971). They consider instead:

a) an affine variety X^n

b) the action of a 1-dimensional torus T^1 on X^n,

c) a fixed point $x \in X$ for this action which is "good": this means that for all $y \in X$, $y \neq x$,

$$x \in \overline{\mathcal{O}(y)}.$$

Under these conditions, they show that

$$Y = X - \{x\}/T$$

is an n-1-dimensional projective variety, and that if X" = [normalization of graph of rational map $X \longrightarrow Y$], then X" has toroidal singularities (in fact, even singularities of the type $\mathbb{A}^n$/cyclic group).

§2. Tits' buildings

Another example of toroidal embeddings and the associated complex is furnished by Tits' buildings (also known as apartments, or flag complexes). These are of 2 types: the "absolute" building attached to a semi-simple group over a general field, based on its parabolic subgroups and built up out of spheres decomposed by the usual Weyl chambers; the "relative" building, due jointly to Brubat and Tits, attached to a semi-simple group over a complete-valued field, based on its parahoric subgroups and built up out of Euclidean spaces decomposed into affine Weyl chambers.

We will deal in this section with the absolute building, and to make the most geometric construction, we will work (as always in these notes so far) over an algebraically closed field k. (If one has a group over a smaller field of definition $k_o \subset k$, it is easy to see that all our construction will lead to schemes rational over k_o, and that by taking $Gal(k/k_o)$-invariant strata, one recovers the building of k_o-rational parabolics.) Fix therefore a semi-simple algebraic group G over k. We propose to construct an embedding:

$$G \underset{\substack{\text{open}\\ \text{dense}}}{\subset} \overline{G}$$

where $\overline{G}$ is an infinite union of affine pieces: to be precise, $\overline{G}$ is a reduced, irreducible and separated scheme, locally of finite type over k. This embedding is to satisfy:

a) $G \subset \overline{G}$ is a toroidal embedding without self-intersection and if G has no center, then $\overline{G}$ is non-singular.

b) The left-action of G on itself extends to an action of G on $\overline{G}$: i.e., there is a morphism

$$\mu: G \times \overline{G} \longrightarrow \overline{G}$$

extending the group law in G,

c) The right action of G on itself extends pointwise but not continuously to an action on $\overline{G}$, i.e., for all $g \in G(k)$, if

$$R_g: G \longrightarrow G$$

is right multiplication by g, then R_g extends to a morphism:

$$\overline{R}_g: \overline{G} \longrightarrow \overline{G},$$

d) For each stratum Y of $\overline{G}-G$,

$$\left\{ g \in G(k) \,\middle|\, \overline{R}_g(Y) = Y \right\}$$

is the set of closed points of a parabolic subgroup P_Y of G, and

$$Y \longmapsto P_Y$$

sets up a bijection between the strata of $\overline{G}-G$ and the parabolics $P \subset G$. This bijection extends to an isomorphism:

$$\left\{ \begin{matrix} \text{complex } \Delta \\ \text{of the embedding} \\ G \subset \overline{G} \end{matrix} \right\} \Big/ \underset{\mathbb{R}_+^*}{\text{homotheties}} \quad \cong \quad \left\{ \begin{matrix} \text{Tits' building} \\ \text{for } G/k \end{matrix} \right\}^*$$

* Cf. for instance Mumford, Geometric Invariant Theory, Ch. 2, §2.

Our construction is similar but not identical to that of Borel and Serre (*Adjonction de coins*, Comptes Rendus, 271 (1970), p. 1156): their construction uses essentially the real ground field and the resulting "geodesic action" on $D = K\backslash G_{\mathbb{R}}$, K = maximal compact subgroup. To visualize the difference between the 2 embeddings of $D = K\backslash G_{\mathbb{R}}$ in the simplest case, take the unit disc $D \subset \mathbb{C}$ ($D = SO(2)\backslash SL(2,\mathbb{R})$) and to construct $\overline{SO(2)\backslash SL(2,\mathbb{R})}$ replace each point on its boundary by an entire real line! We may define convergence of interior points to boundary points in 2 ways, with the following shaded areas as typical open sets meeting the boundary line at $1 \in \partial D$ in an interval:

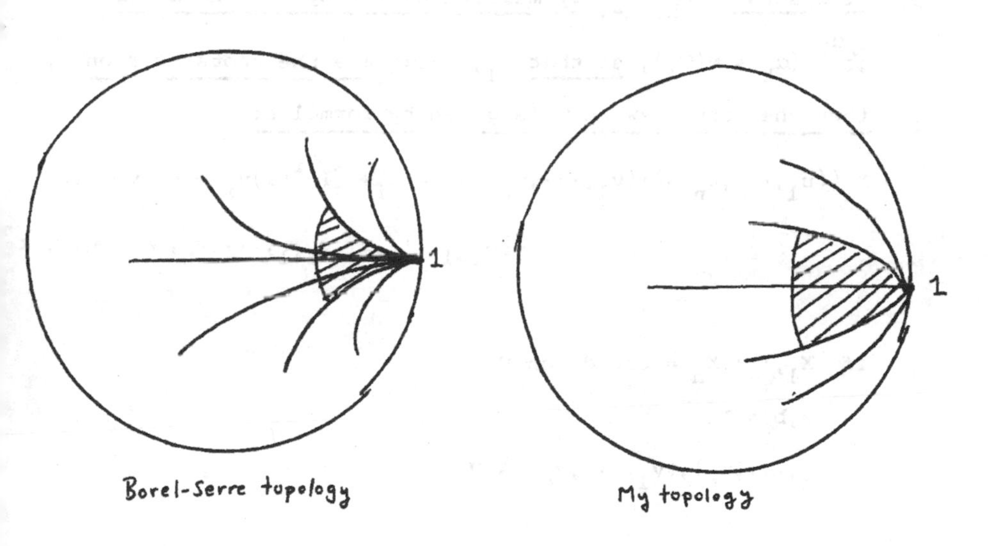

Before embarking on the construction of $\overline{G}$, we must prove several preliminary results on embeddings of solvable groups. For these in turn I need a structure theorem for solvable groups which seems on the verge of being "well-known", but which I cannot locate in the literature:

Theorem: *Let G be a solvable group. Then:*

1) $G \cong U \times T$ (*semi-direct product*), *T a torus, U unipotent,*

2) $U \underset{\text{as schemes}}{\cong} \mathbb{A}^n$, *in such a way that for all* k,

$$U_k = \mathbb{A}^k \times \overbrace{(0,\cdots,0)}^{n-k}$$

is a normal subgroup of G and projection on the kth coordinate defines a group isomorphism

$$U_k/U_{k-1} \cong \mathbb{G}_a,$$

3) *If T acts on U_k/U_{k-1} by multiplication by the character $\mathfrak{X}^{\alpha_k}$ ($\alpha_k \in M(T)$), so that $\alpha_1,\cdots,\alpha_n$ are the roots of T on U, then the group law in G is given by formulae:*

$$X_i((u_1,\cdots,u_n,a)\cdot(v_1,\cdots,v_n,b)) = u_i + \mathfrak{X}^{\alpha_i}(a)v_i + \text{polyn. in } [u_{i+1},\cdots,u_n,v_{i+1},\cdots,v_n,\mathfrak{X}^{\alpha_i}(a),\cdots,\mathfrak{X}^{\alpha_n}$$

if $X_1,\cdots,X_n$ = coord. on U

$a,b \in T$

$(u_1,\ldots,u_n),(v_1,\ldots,v_n) \in U$.

Proof: (1) and (2) are standard (cf. Borel, Linear Algebraic Groups, Benjamin, §10). It is (3) that I can't locate. It suffices to prove the formula for $i = 1$, since the general case is just this case applied to G/U_{i-1}. Also, the formula

$$X_1((u_1,\cdots,u_n,a)\cdot(v_1,\cdots,v_n,b)) = u_1 + \mathfrak{X}^{\alpha_1}(a)\cdot v_1 + \text{polyn. in } [u\text{'s}, v\text{'s and } \mathfrak{X}^{\beta}(a), \text{ all } \beta]$$

is clear, e.g., via Borel, (10.10). My small strengthening results from:

Lemma: Let a torus T act on an affine variety X via $\sigma: T\times X \longrightarrow X$ and let $x \in X$ be fixed. Let $\mathfrak{X}^{\alpha_i}$, $1 \leq i \leq n$, be the characters in the action of T on $T^*_{x,X}$. Then

$$\sigma^*\Gamma(\mathcal{O}_X) \subset k[\mathfrak{X}^{\alpha_1},\cdots,\mathfrak{X}^{\alpha_n}] \otimes \Gamma(\mathcal{O}_X).$$

Proof: Since we can diagonalize the action of T on $\Gamma(\mathcal{O}_X)$, it suffices to prove that if $f \in \Gamma(\mathcal{O}_X)$ and $\sigma^*f = \mathfrak{X}^{\alpha}\otimes f$, then $\alpha \in$ [semi-group gen. by $\alpha_1,\cdots,\alpha_n$]. But $f \in \mathcal{O}_{x,X}$ is not 0, so suppose $f \in m^r_{x,X} - m^{r+1}_{x,X}$. Then $\bar{f} \in m^r_{x,X}/m^{r+1}_{x,X}$ also satisfies $\sigma^*\bar{f} = \mathfrak{X}^{\alpha}\otimes\bar{f}$. Since the T-representation space $m^r_{x,X}/m^{r+1}_{x,X}$ is a quotient of the rth symmetric power of the T-representation space $m_{x,X}/m^2_{x,X} = T^*_{x,X}$, it follows that α is a positive linear combination of the α_i.

Apply the lemma with $X = U$, σ = conjugation, x = identity.

QED for Theorem

We will not try to classify all embeddings of a solvable group $G = U \times T$, but consider simply:

$$Y_\sigma = G \times^T X_\sigma$$

where σ = polyhedral cone in $N_{\mathbb{R}}(T)$

$T \subset X_\sigma$ the associated T-embedding

$G \times^T X_\sigma$ means $G \times X_\sigma$ modulo the equivalence relation $(gt,x) \sim (g,tx)$, $t \in T$.

In fact

$$Y_\sigma = (U \times T) \times^T X_\sigma \cong U \times X_\sigma$$

hence Y_σ exists (quotients don't automatically exist in the category of varieties) and is affine. Moreover

$$\begin{array}{c} Y_\sigma = G \times^T X_\sigma \\ \cup \text{ open subset} \\ G = G \times^T T \end{array}$$

so this is an affine embedding of G. Then I claim:

a) <u>the left action of</u> G <u>on itself extends to an action of</u> G <u>on</u> Y_σ,

b) <u>if</u> $\alpha_1, \cdots, \alpha_n$ <u>are the roots for the action of</u> T <u>on</u> U, <u>and</u> $\rho = \left\{ x \in N_{\mathbb{R}}(T) \,\middle|\, \langle \alpha_i, x \rangle \geq 0, \text{ all } i \right\}$, <u>then the following are equivalent</u>:

(i) <u>the right action of</u> G <u>on itself extends to an action of</u> G <u>on</u> Y_σ,

(ii) $\sigma \subseteq \rho$.

If these hold, then in fact:

(iii) Y_σ is an algebraic semi-group: $\exists\, Y_\sigma \times Y_\sigma \longrightarrow Y_\sigma$, extending the group law on G,

(iv) Y_σ is independent of the choice of T in G, i.e., if aTa^{-1} is any other maximal torus, and
$Y'_\sigma = G \times^{(aTa^{-1})} X_\sigma = [G\times X_\sigma \text{ modulo} (gata^{-1},x) \sim (g,tx)]$
then there is an isomorphism χ:

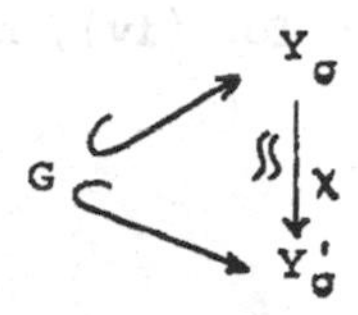

Proof: (a) is immediate. To prove (b), note that the inclusion $G \subset Y_\sigma$ is the same as $U\times T \subset U\times X_\sigma$, hence on the ring level, corresponds to:

$$k[\cdots,\mathfrak{X}^\alpha,\cdots;X_1;\cdots;X_n]_{\alpha\in M(T)} \supset \overbrace{k[\cdots,\mathfrak{X}^\alpha,\cdots;X_1,\cdots,X_n]_{\alpha\in\sigma\cap M(T)}}^{\text{call this } R_\sigma}.$$

(Notation as in Structure Theorem.) If $\mu: G\times G \longrightarrow G$ is the group law, then μ extends to $\bar{\mu}: Y_\sigma \times G \longrightarrow Y_\sigma$ if and only if

$$\mu^*(X_i) \in R_\sigma \otimes \Gamma(\mathcal{O}_G)$$

(since $\mu^*(\mathfrak{X}^\alpha) = \mathfrak{X}^\alpha \otimes \mathfrak{X}^\alpha$, this element is always in $R_\sigma \otimes \Gamma(\mathcal{O}_G)$).
Our structure theorem says

$$\mu^* X_i = X_i \otimes 1 + \mathfrak{X}^{\alpha_i} \otimes X_i + \left[\begin{array}{l}\text{polyn. in } X_j\otimes 1,\ 1\otimes X_j,\ j > i \\ \quad \text{and } \ \mathfrak{X}^{\alpha_j}\otimes 1,\ j \geq i\end{array}\right]$$

Therefore $\mu^*(X_i) \in R_\sigma \otimes \Gamma(\mathcal{O}_G)$ for all i if and only if $\mathfrak{X}^{\alpha_i} \in R_\sigma$, all i, hence if and only if $\alpha_i \geq 0$ on σ, all i. Since this is equivalent to $\rho \supseteq \sigma$, (i) and (ii) are equivalent. Moreover if $\mathfrak{X}^{\alpha_i} \in R_\sigma$, then we see that

$$\mu^*(X_i) \in R_\sigma \otimes R_\sigma,$$

hence $\mu^*(R_\sigma) \subset R_\sigma \otimes R_\sigma$. This defines a semi-group morphism $\mu: Y_\sigma \times Y_\sigma \longrightarrow Y_\sigma$, and proves (iii). As for (iv), note that we have a commutative diagram:

$$\begin{array}{ccc} G & \subset & G \times^{(aTa^{-1})} X_\sigma \\ \text{right mult. } R_a \text{ by a} \downarrow \wr & & \phi \downarrow \wr \\ G & \subset & G \times^T Y_\sigma \end{array}$$

where $\phi(g,x) = (ga,x)$.

But assuming (i), there is also an isomorphism ψ:

$$\begin{array}{ccc} G & \subset & G \times^T Y_\sigma \\ R_a \downarrow & & \wr\downarrow \psi \\ G & \subset & G \times^T Y_\sigma \end{array}$$

Therefore $\chi = \phi^{-1} \cdot \psi$ has the required property. QED

Corollary: Given any solvable group G, there is a canonical* affine embedding $G \subset \overline{G}$ (essentially the biggest which is a semi-group):

---

*N.B. Actually there are 2 such: if we split G into T×U instead of U×T, we would get a different $\overline{G} = X_\rho \times^T G$!

<u>let</u> U = <u>unipotent radical of</u> G

<u>let</u> $T = G/U$, $\alpha_i \in M(T)$ <u>the roots of</u> T <u>in</u> U

<u>let</u> $\rho \subset N_{\mathbb{R}}(T)$ <u>be</u> $\{x \mid \langle\alpha_i, x\rangle \geq 0, \text{ all } i\}$,

<u>split</u> $G = U \times T$

<u>form</u> $\overline{G} = G \times^T X_\rho$.

<u>Only the</u> 4th <u>step is non-canonical and</u> $\overline{G}$ <u>does not depend on the choice of the splitting.</u>

Now return to a semi-simple group G. For every Borel subgroup $B \subset G$, form $\overline{B}$, the above canonical embedding. We shall define:

$$\overline{G} = \bigcup_{\substack{\text{Borel} \\ \text{subgps.} \\ B}} G \times^B \overline{B}$$

where the glueing is the unique one so that $^{(1)}$ all $G \times^B \overline{B}$'s are identified at least on their common open set $G \cong G \times^B B \subset G \times^B \overline{B}$, and so that $^{(2)}\overline{G}$ is separated and $G \times^B \overline{B} \subset \overline{G}$ is an open subset. Now first of all, $G \times^B \overline{B}$ exists: in fact consider the projection:

$$p: G \longrightarrow G/B .$$

This makes G into a principle fibre bundle over G/B, with structure group B, locally trivial in the Zariski topology over the projective variety G/B. Since $G \times^B \overline{B}$ is just the associated fibre bundle with fibre $\overline{B}$, it exists. Secondly to check that the glueing can be done with Properties (1), (2) and (3), we must check that for any 2 Borels B_1, B_2, if

$$i_\ell : G \longrightarrow G \times^{B_\ell}(\overline{B}_\ell)$$

is the canonical open immersion, then

$$\text{Closure Im}\left[(i_1,i_2) : G \longrightarrow (G\times^{B_1}\overline{B}_1)\times(G\times^{B_2}\overline{B}_2)\right]$$

is the graph of an isomorphism between open sets $U_{12};U_{21}$:

$$\begin{array}{ccccc} i_1(G) & \subset U_{12} & \subset & G \times^{B_1}(\overline{B}_1) \\ \wr\| & \wr\| & & \\ i_2(G) & \subset U_{21} & \subset & G \times^{B_2}(\overline{B}_2) \ . \end{array}$$

But $B_1 \cap B_2$ contains a maximal torus T, so by construction:

$$G \times^{B_1}(\overline{B}_1) \cong G \times^{B_1}[B_1 \times^T T_{\rho_1}] = G \times^T T_{\rho_1}$$

$$G \times^{B_2}(\overline{B}_2) \cong G \times^{B_2}[B_2 \times^T T_{\rho_2}] = G \times^T T_{\rho_2}$$

where $\rho_\ell \subset N_{\mathbb{R}}(T)$ is nothing but the positive Weyl chamber associated to the Borel subgroup B_ℓ containing T. But therefore $\rho_1 \cap \rho_2$ is a face of ρ_1 and of ρ_2, hence $T_{\rho_1 \cap \rho_2}$ is an open subset of both T_{ρ_1} and T_{ρ_2} whose image in $T_{\rho_1} \times T_{\rho_2}$ is closed. Therefore

$$U_{12} = \text{Im}\left[G \times^T(T_{\rho_1 \cap \rho_2}) \longrightarrow G \times^T(T_{\rho_1})\right]$$

$$U_{21} = \text{Im}\left[G \times^T(T_{\rho_1 \cap \rho_2}) \longrightarrow G \times^T(T_{\rho_2})\right]$$

are the needed open subsets. More succinctly, we may glue $G\times^{B_1}(\overline{B}_1)$ and $G\times^{B_2}(\overline{B}_2)$ together to form $G\times^T(T_{\{\rho_1,\rho_2\}})$. In fact, for any fixed maximal torus T, it follows that

$$\bigcup_{\substack{\text{Borels'} \\ B\supset T}} (G \times^B \overline{B}) \cong G \times^T \overline{T}$$

where $\overline{T} = \left\{\begin{array}{l}\text{complete variety containing T as open set} \\ \text{associated to the decomposition of the whole} \\ \text{vector space } N_{\mathbb{R}}(T) \text{ into Weyl chambers}\end{array}\right\}.$

This completes the construction of $\overline{G}$.

Let's check that $\overline{G}$ has all the properties asserted at the beginning of this section:

Proof of a) locally $\overline{G}$ is isomorphic to $G\times^T X_\rho$, hence locally it is isomorphic to $G/T \times X_\rho$. Since $T \subset X_\rho$ is even a torus embedding, $G \subset \overline{G}$ is toroidal without self-intersection. If G has no center, the roots generate M(T) and for each $B \supset T$, among the roots occurring in Lie(B), one can find so-called "fundamental roots" $\alpha_1,\cdots,\alpha_\ell$ such that a) they form a basis of M(T), b) all roots in Lie(B) are positive linear combinations of the α_i. It follows that the chamber ρ defined by B is:

$$\rho = \{x \mid \langle\alpha_i,x\rangle \geq 0,\ 1 \leq i \leq \ell\}$$

$$= \text{span of the basis } e_1,\cdots,e_\ell \text{ of } N(T) \text{ dual to } \{\alpha_i\}.$$

Thus X_ρ is non-singular, hence so is $\overline{G}$.

Proof of b) is clear.

Proof of c) for any closed point $g \in G$, the conjugation $a \longmapsto gag^{-1}$ extends to isomorphisms:

$$C_G: \; B \xrightarrow{\approx} gBg^{-1}$$

$$\overline{C}_g: \; \overline{B} \xrightarrow{\approx} \overline{gBg^{-1}}$$

for all Borel subgroups B. Then we get a diagram:

$$\begin{array}{ccc} G & \subset & G \times^B \overline{B} \\ \wr\wr \big\downarrow R_{g^{-1}} & & \wr\wr \big\downarrow R_{g^{-1}} \times \overline{C}_g \\ G & \subset & G \times^{gBg^{-1}} (\overline{gBg^{-1}}) \end{array}$$

from which it follows immediately that $R_{g^{-1}}: G \longrightarrow G$ extends to an isomorphism of $\overline{G}$ with $\overline{G}$ which permutes the varous open pieces $U_B = G \times^B \overline{B}$ by conjugation.

Proof of d) Fix $B \subset G$ a Borel subgroup and $T \subset B$ a maximal torus. Consider first the strata Y in the middle open piece:

$$G \subset G \times^B \overline{B} \cong G \times^T T_\rho \subset G \times^T \overline{T}.$$

Since B acts by a morphism on the right, and B is connected, B must map each stratum Y into itself. Now the strata in $G \times^T T_\rho - G$ correspond 1-1 with the strata in $T_\rho - T$, hence with the faces of ρ: but by the standard theory, these correspond 1-1 with the parabolics $P \supset B$, i.e., $P \rightleftarrows$ face σ if

a) Lie P is spanned by roots α with $\alpha \geq 0$ on σ

or b) P generated by B and those $w \in N(T)$ such that

$$c_w(\sigma) = \sigma \quad (c_w = \text{conjugation by } w).$$

But if in this way Y corresponds to σ, hence to P, then P = stabilizer of Y: because $B \subset Stab(Y) \subset G$ and any such group is generated by B and a subset of N(T). And if $w \in N(T)$, then w acts on $G \times^T \bar{T}$ and $w(Y) = Y$ iff $w(\sigma) = \sigma$ iff $w \in P$. Therefore we have a map:

$$(\text{Strata in } \bar{G}\text{-}G) \longrightarrow (\text{Parabolic subgroups of } G)$$

by $Y \longmapsto Stab(Y)$. Since every parabolic P containing B is associated to some face of ρ, the map is surjective. As for injective, take 2 strata $Y_\ell \subset G \times^{B_\ell} (\overline{B_\ell})$: choose $T \subset B_1 \cap B_2$. Then $Y_\ell \subset [G \times^T \bar{T}\text{-}G]$ correspond to 2 simplices σ_ℓ in the Weyl chamber decomposition of $N_{\mathbb{R}}(T)$. If $Y_1 \neq Y_2$, then $\sigma_1 \neq \sigma_2$, hence the corresponding parabolics are distinct.

Now in particular the codim. 1 strata of $\bar{G}$-G correspond to the maximal parabolics $P \subset G$: so the graph of $\bar{G}$-G mod $\mathbb{R}^*_+$ can be obtained as follows — take a vertex v(P) for every maximal parabolic; for any $P_1,\cdots,P_k$, fill in $v(P_1),\cdots,v(P_k)$ by a (k-1)-simplex if and only if $P_1 \cap\cdots\cap P_k$ contains a Borel subgroup B. But this is Tits' building.

§3. Extensions to schemes over R

Up to this point, all our theory has dealt with varieties — i.e., reduced and irreducible schemes of finite type over an algebraically closed field k — or at least with schemes locally of finite type. But much of our theory works equally well for schemes locally of finite type over discrete valuation rings R. The purpose of this section is to quickly extend our results to this case. For this whole section, we fix the notations:

R = discrete valuation ring

m = maximal ideal of R, $\pi \in m-m^2$ a generator of m

k = R/m

K = quotient field of R

S = Spec R

η = generic point of S

o = closed point of S

$\forall$ schemes X/S, X_η = fibre over η

X_o = fibre over o.

(I.) Suppose T is a split torus over R, i.e., $T \cong (\mathbb{G}_m^n$ over R), we can consider torus embeddings:

$$(*)\quad \left\{\begin{array}{l} T_\eta \underset{\text{open}}{\subset} X \\ \searrow \quad \swarrow \\ \quad S \\ \text{X irreducible normal and of finite type over R,} \\ \text{with T acting on X so as to extend the translation} \\ \text{action of } T_\eta \text{ on itself.} \end{array}\right.$$

We consider:

$M(T)$ = character group of T

$N(T) = \mathrm{Hom}(M(T),\mathbb{Z})$

$N_{\mathbb{R}}(T) = N(T)\otimes\mathbb{R}$

$\tilde{M}(T) = M(T)\times\mathbb{Z}$

$\tilde{N}(T) = N(T)\times\mathbb{Z}$

$\tilde{N}_{\mathbb{R}}(T) = N_{\mathbb{R}}(T)\times\mathbb{R}$

Then for all rational polyhedral cones

$$\sigma \subset N_{\mathbb{R}}(T) \times \mathbb{R}_+ ,\ \sigma \not\supset \text{linear subspace of } N_{\mathbb{R}}(T),$$

let

$$X_\sigma = \mathrm{Spec}\, R[\cdots,\pi^k\cdot\mathfrak{X}^\alpha,\cdots]_{(\alpha,k)\in\tilde{M}(T)\cap\check{\sigma}} .$$

As a simple example, the embedding $T_\eta \subset T$ is defined by the cone $\sigma = (0)\times\mathbb{R}_+$. Then exactly as in Ch. I, one proves:

a) X_σ is normal of finite type over R with T acting on it and with T_η as an open set. $\sigma \longmapsto X_\sigma$ sets up a bijection between these cones σ and such affine embeddings of T.

b) Moreover, there is a bijection between orbits of T_η in $(X_\sigma)_\eta$ and faces of σ in $N_{\mathbb{R}}(T)\times(0)$; and orbits of T_0 in

$(X_\sigma)_0$ and faces of σ not contained in $N_{\mathbb{R}}(T)\times(0)$; under this bijection $\sigma_1 \subseteq \sigma_2$ iff $\overline{(\text{orbit of } \sigma_1)} \supseteq (\text{orbit of } \sigma_2)$ and:

$$\dim \sigma = (\text{codim in } X_\sigma \text{ of corresponding orbit}).$$

c) There is a diagram

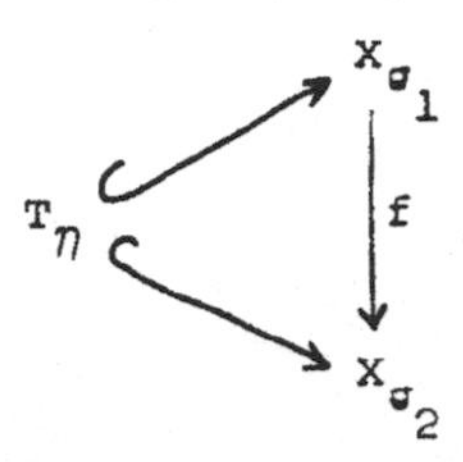

iff $\sigma_1 \subseteq \sigma_2$, and f is an open immersion iff σ_1 = face of σ_2.

d) X_σ is a regular scheme iff $\sigma \cap (N(T)\times\mathbb{Z})$ can be generated by a subset of a basis of $N(T)\times\mathbb{Z}$.

e) If $T_\eta \subset X$ is any torus embedding as in (*), then X is covered by T-invariant affines and hence such X's correspond bijectively to "f.r.p.p. decompositions of $\widetilde{N}_{\mathbb{R}}(T)$" $\{\sigma_\alpha\}$ (via $X = \bigcup_\alpha X_{\sigma_\alpha}$).

e') If X and $\{\sigma_\alpha\}$ correspond, then there are bijections between the set of T_η-orbits in X_η and the set of $\sigma_\alpha \subset N_{\mathbb{R}}(T)\times(0)$; and the set of T_0-orbits in X_0 and the set of $\sigma_\alpha \not\subset N_{\mathbb{R}}(T)\times(0)$. For instance, the smaller toroidal embedding $T_\eta \subset X_\eta$ corresponds to the induced decomposition of $N_{\mathbb{R}}(T)\times(0)$;

and the components E_α of X_o correspond to 1-dimensional $\sigma_\alpha = \mathbb{R}^+\cdot(v_\alpha, k_\alpha)$, $v_\alpha \in N(T)$, $k_\alpha \geq 1$. Here if (v_α, k_α) is primitive in $N(T)\times\mathbb{Z}$, then k_α = order of vanishing of π along E_α.

f) There is a diagram

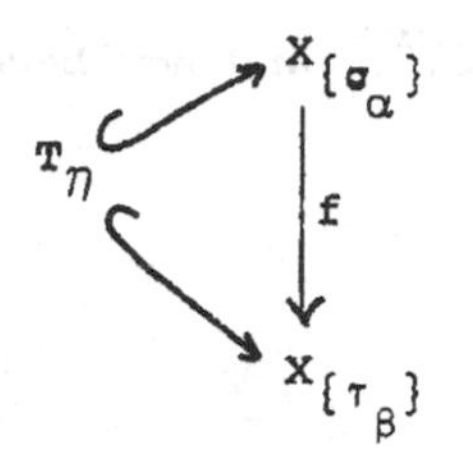

iff for all α, $\sigma_\alpha \subseteq \tau_\beta$ for some β.

g) $X_{\{\sigma_\alpha\}} \longrightarrow X_{\{\tau_\beta\}}$ is proper iff $\bigcup \sigma_\alpha = \bigcup \tau_\beta$;

$X_{\{\sigma_\alpha\}} \longrightarrow S$ is proper iff $\bigcup \sigma_\alpha = \tilde{N}_{\mathbb{R}}(T)$.

h) T-invariant complete sheaves of fractional ideals $\mathfrak{J}$ on such X are classified by piecewise-linear locally convex functions f = ord $\mathfrak{J}$ on $\bigcup \sigma_\alpha$, exactly as in Th. 9.

i) $B_{\mathfrak{J}}(X_{\{\sigma_\alpha\}}) = X_{\{\tau_\beta\}}$, where the τ_β are the biggest sub-polyhedra of σ_α where f is linear.

j) For any torus embedding $T_\eta \subset X$, there exists an $\mathfrak{J}$ as in (h) such that $B_{\mathfrak{J}}(X)$ is regular.

k) An invertible sheaf $\mathfrak{J}$ on X is ample iff f = ord $\mathfrak{J}$ is strictly convex on $\bigcup \sigma_\alpha$.

The only really new point to observe here is this: if V is an R-module on which T acts, then by the diagonalizability of the action of T, we can write

$$V = \bigoplus_{\alpha \in M(T)} V_\alpha$$

where V_α is an R-submodule on which T acts through the character $\mathfrak{X}^\alpha$. Moreover if $V \subset K(T_\eta)$, then $V_\alpha \subset K\cdot\mathfrak{X}^\alpha$. And now because R is a discrete valuation ring, then

$$\begin{aligned} \text{either } V_\alpha &= (0) \\ \text{or } V_\alpha &= K\cdot\mathfrak{X}^\alpha \\ \text{or } V_\alpha &= R\cdot\pi^k\cdot\mathfrak{X}^\alpha, \qquad \text{some } k \in \mathbb{Z}! \end{aligned}$$

This has the effect that there might just as well have been a torus acting "horizontally" in the S-direction as well as vertically over S: just as in §1 of this Chapter, you never miss that final dimension of T. The set of monomials $\pi^k\cdot\mathfrak{X}^\alpha$ take the place of the characters $\mathfrak{X}^\alpha$ alone. In particular, this proves that a finitely generated R-algebra $A \subset K(T_\eta)$ on which T acts by translation is generated by a finite set of monomials $\pi^k\cdot\mathfrak{X}^\alpha$. This would fail completely if the base S were Spec R, dim R $\geq$ 2.

(II.) Suppose now that X is any irreducible normal scheme, of finite type over R and that $U \subset X_\eta$ is an open set smooth over K. Then we say that $U \subset X$ is a toroidal embedding without self-intersection if for all $x \in X$, there is

i) an open neighborhood $U_x \subset X$ of x,

ii) a torus embedding $T_\eta \subset Z_{\rho(x)}$

iii) an étale morphism $\pi: U_x \longrightarrow Z_{\rho(x)}$ such that $U \cap U_x = \pi^{-1}(T_\eta)$.

This is superficially quite different from our definition in Ch. II. In fact, though, it is both equivalent when we are working over an algebraically ground field k*, and it seems in fact easier to work with. Because the residue fields $\mathbb{k}(x)$ even of closed points $x \in X$ may not be algebraically closed, this new definition is forced on us.

We then analyze such embeddings very much as in Chapter II:

a) let $E_1, \cdots, E_N$ = components of X-U. Then one finds

$\bigcap_{i \in I} E_i$ normal,

$\bigcap_{i \in I} E_i - \bigcup_{i \notin I} E_i$ regular.

---

* Obviously the definition with an étale π implies the former definition: $\exists\ \phi: \hat{\mathcal{O}}_{x,X} \xrightarrow{\approx} \hat{\mathcal{O}}_{t,Z_\rho}$ with $I(X-U)\cdot\hat{\mathcal{O}}_{x,X} \xrightarrow{\approx} I(Z_\rho - T)\cdot\hat{\mathcal{O}}_{t,Z_\rho}$. Conversely assuming for simplicity that t is a fixed point for T, ϕ sets up an isomorphism $M(T) \xrightarrow{\approx}$ [Gp. of C. div. at x, supported by X-U]. Under this isomorphism, choose local equations f^α for the divisor corresponding to $\mathfrak{X}^\alpha$ so that $f^{\alpha+\beta} = f^\alpha \cdot f^\beta$. Define π: (Nghd. of x)$\to Z_\rho$ by $\pi^*(\mathfrak{X}^\alpha) = f^\alpha$ whenever $\alpha \geq 0$ on ρ.

Define the strata to be the components of the latter.

b) for each stratum Y, define

M^Y = gp. of C-divisors on Star (Y), supposed on Star (Y)-U

M^Y_+ = subsemi-group of effective divisors

$N^Y = \mathrm{Hom}(M^Y, \mathbb{Z})$

$N^Y_{\mathbb{R}} = N^Y \otimes \mathbb{R}$

$\sigma^Y = \left\{ x \in N^Y_{\mathbb{R}} \,\middle|\, x \geq 0 \text{ on } M^Y_+ \right\}$

c) Define Δ = [the conical complex $\bigcup \sigma^Y$], where for each stratum $Z \subset$ Star Y, we identify σ^Z with a suitable face of σ^Y. By restriction of the linear functions M^Y, define an integral structure on Δ.

There is no really satisfactory analog for $R.S.^U(X)$ and ord: $R.S.^U(X) \longrightarrow \Delta$ in the general case. If, however, we consider the special case $U = X_\eta$, then one can consider

$\tilde{R}$ = integral closure of the completion $\hat{R}$ of R in an algebraic closure of the quotient field of $\hat{R}$

R.S.(X) = [set of R-morphisms Spec $\tilde{R} \longrightarrow X$].

Note that $\tilde{R}$ is a valuation ring and if we identify the (isomorphic) value groups of R and $\hat{R}$ with $\mathbb{Z}$, then we get a canonical identification of the value group of $\tilde{R}$ with $\mathbb{Q}$. For each $\phi \in$ R.S.(X), if $\phi(\tilde{0}) \in$ stratum Y, we get a pairing

$$\langle \phi, D \rangle = v_{\tilde{R}}\left(\phi^*\binom{\text{local eq.}}{\text{d of D}}\right) \in \mathbb{Q},$$

hence a point ord $\phi \in \sigma^Y$. This defines a map

$$\text{ord}: \text{R.S.}(X) \longrightarrow \Delta.$$

In fact, since $U = X_\eta$, the divisor (π) has support exactly X-U and π defines a global function $\ell_\pi: \Delta \longrightarrow \mathbb{R}_+$ which is 0 only at $0 \in \Delta$. Thus we have also a canonical choice of a compact polyhedral complex $\Delta_o \subset \Delta$:

$$\Delta_o = \bigcup_Y \left\{ x \in N_{\mathbb{R}}^Y \,\middle|\, x \geq 0 \text{ on } M_+^Y, \langle x,(\pi)\rangle = 1 \right\}$$

$$= \left\{ x \in \Delta \,\middle|\, \ell_\pi(x) = 1 \right\}.$$

Since $v_{\tilde{R}}(\phi^*\pi) = 1$ for all $\phi \in$ R.S.(X), ord(ϕ) in fact lies in Δ_o for all ϕ.

Next the theorems in Ch. II, §2, extend to this case. (The proofs here must be modified much more than those in Ch. I, but I think, in fact, they are easier using the new definition of toroidal embedding):

d) let $U \subset X$ be any toroidal embedding without self-intersection. Then for all f.r.p.p. decompositions Δ' of the conical complex Δ associated to (X,U), we get a new toroidal embedding $U \subset X'$ and a morphism f

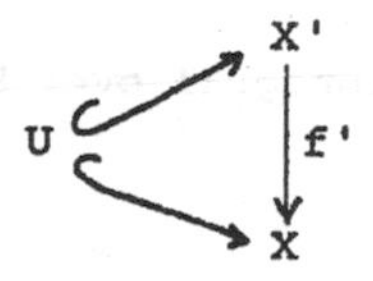

such that Δ' is the complex of (X',U) and f defines the given piecewise-linear map $\Delta' \longrightarrow \Delta$.

e) For each piecewise linear locally convex function f on Δ (as in Th. 9*), we get a coherent sheaf $\mathfrak{J}_f$ of fractional ideals on X, and $B_{\mathfrak{J}_f}(X)$ is the modification of X associated to the decomposition of Δ into the biggest subpolyhedra on which f is linear.

f) If $U \subset X$ is a toroidal embedding without self-intersection and $U = X_\eta$, then

i) π vanishes to 1st order on all components of X_o iff the vertices of Δ_o are integral (with respect to the integral structure defined in (c)),

ii) X is regular and π vanishes to 1st order on all components of X_o iff the vertices of Δ_o are integral and the polyhedra $\sigma^Y \cap \Delta^o$ of Δ^o are all simplices of volume $1/r_Y!$ (if $\dim \sigma^Y = r_Y+1$).

We can then proceed to prove:

Semi-stable reduction theorem II: Let R be a discrete valuation ring with char.$(R/(\pi)) = 0$ (n.b. such a ring is always "excellent": cf. EGA IV, (7.8)). Let $f: X \longrightarrow S = \text{Spec } R$ be a morphism of finite type from a normal scheme X to S, smooth over $\eta \in S$. Then there is a discrete valuation ring $R' \supset R$, integral over R, and a proper morphism P as follows:

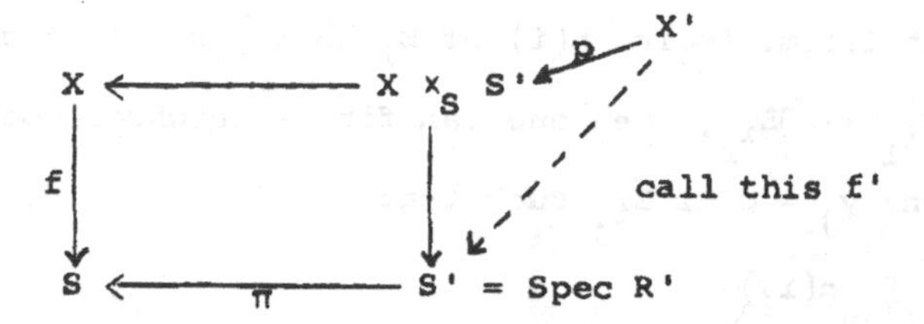

such that

a) p is an isomorphism over $\eta' \in S'$,

b) p is projective; in fact, p is obtained by blowing up a sheaf of ideals $\mathcal{J}$ with $\mathcal{J} = \mathcal{O}_{X \times_S S'}$ over η',

c) X' is regular and the fibre $f'^{-1}(o')$ is reduced with regular components crossing normally.

The proof is essentially the same as in the algebraic case: the construction in Ch. III is used without change and Hironaka's resolution theorems apply to schemes such as X because they are excellent (note the definition on p. 161 of Hironaka's fundamental paper, Annals of Math., 79 (1964) and EGA IV (7.8.3)). The only point to mention is, after reduction using Hironaka to the case X regular, $(X_o)_{red}$ with regular components E_i crossing normally, and construction of

$$R_d = R[\sqrt[d]{\pi}]$$

$$S_d = \text{Spec } R_d$$

$$X_d = \text{normalization of } X \times_S S_d,$$

why $(X_d)_\eta \subset X_d$ is a toroidal embedding without self-intersection for all d such that l.c.m. $\left(\text{mult. } n(i) \text{ of } E_i \text{ in } X_o\right) \Big| d$? In fact, if $x \in (X)_o$ lies on $E_{i_1} \cap \cdots \cap E_{i_r}$, then one can find a neighborhood U_x of x and local equations $y_j = 0$ of E_{i_j} such that

$$\pi = u \cdot \prod_{j=1}^{r} y_j^{n(i_j)}$$

$$u \in \Gamma(U_x, \mathcal{O}_X^*).$$

Moreover, since $\mathcal{O}_{x,X}$ is regular and the E_{i_j} meet transversely, we can choose $y_{r+1}, \cdots, y_s, y_{s+1}, \cdots, y_n \in \mathcal{O}_{x,X}$ so that $\bar{y}_1, \cdots, \bar{y}_s \in m_x/m_x^2$ are an $\mathcal{O}_x/m_x$-basis and $\bar{y}_{s+1}, \cdots, \bar{y}_n \in \mathcal{O}_x/m_x$ are a transcendence basis over $k = R/(\pi)$. Let $e = \text{g.c.d. } (n(i_1), \cdots, n(i_r))$ and choose $a(j) \in \mathbb{Z}$ so that

$$e = \sum_{j=1}^{r} a(j) \cdot n(i_j).$$

Let $\tau = \sqrt[d]{\pi}$, and let V_x be the inverse image of U_x in X_d. Then in V_x,

$$\left(\tau^{d/e}\right)^e = u \cdot \left[\prod_{j=1}^{r} y_j^{n(i_j)/e}\right]^e,$$

hence because X_d is normal u has an eth root v in $\Gamma(V_x, \mathcal{O}_{X_d})$ such that

$$\tau^{d/e} = v \cdot \prod_{j=1}^{r} y_j^{n(i_j)/e}$$

hence

$$\tau^{d/e} = \prod_{j=1}^{r} (y_j')^{n(i_j)/e}$$

where

$$y_j' = v^{a(j)} \cdot y_j.$$

Now define:

$$Z_d = \text{normalization of Spec } R[x_1,\cdots,x_n]\Big/\Big(\tau^{d/e}-\prod_{j=1}^{r} z_j^{n(i_j)/e}\Big)$$

$$f: V_x \longrightarrow Z_d \text{ by } f^*(\tau) = \tau,\ f^*(z_j) = y_j'$$

$$T = \text{the torus Spec } R[x_1,x_1^{-1},\cdots,x_n,x_n^{-1}]\Big/\Big(1-\prod_{j=1}^{r} x_j^{n(i_j)/e}\Big).$$

So T acts on Z_d by multiplication of each z_i by x_i; and using the orbit of the K-rational point $z_j = \tau^{d.a(j)/e} \in Z_d$, we can define an isomorphism of T_η with a Zariski open subset of $(Z_d)_\eta$. This gives a torus embedding $T_\eta \subset Z_d$. Finally, f is étale near every point over x: to prove this, factor $V_x \longrightarrow U_x$ and "descend" f as follows:

$$\begin{array}{ccc} V_x & \xrightarrow{\quad f \quad} & Z_d \\ \downarrow & & \downarrow \\ U_x' = \left[\begin{array}{l}\text{covering of } U_x \\ \text{given by } \sqrt[e]{u} = v\end{array}\right] & \xrightarrow{\quad f_o \quad} & \text{Spec } R[z_1,\cdots,z_n]\Big/\Big(\pi-\prod_{j=1}^{r} z_j^{n(i_j)}\Big) = Z_d' \\ \downarrow & & \\ U_x & & \end{array}$$

where $f_o^*(\pi) = \pi$, $f_o^*(z_j) = v^{a(j)}.y_j$. Then U_x' and Z_d' are regular (for U_x', this is because u is nowhere 0, and char k $\nmid$ e!) and for all x' over x, one checks immediately that the scheme $f_o^{-1}(f_o(x'))$ is $\{x'\}$ with reduced structure (of course the residue field extension $\mathbb{k}(x') \supset \mathbb{k}(f_o(x'))$ is separable), hence f_o is étale at x'. But

$V_x \cong Z_d \times_{Z'_d} U'_x$, so f is étale at all points over x. The rest of the proof repeats closely the proof given in Ch. II, §3.

The final topic we wish to treat is the realization via toroidal embeddings of the relative case of Tits' buildings. These are quite recent constructions, announced by Bruhat and Tits in several Comptes Rendus Notes (tome 263, pp. 598-601; p. 766-768; p. 822-825; p. 867-869), but only now being fully published (Publ. I.H.E.S., vol. 41, contains Ch. 1) and on such an ample and extended scale that it is reminiscent of Grothendieck. However, as with the absolute case, we will stick with the question of giving an interpretation only to most "geometric" situations: in this case, that of a split semi-simple and simply connected group scheme G over S. For this, the necessary background is contained in the much more digestible paper of Iwahori and Matsumoto ("On some Bruhat decomposition...", Publ. I.H.E.S., vol. 25). We recall the following basic objects:

i) $T \subset G$ a maximal split torus over S

ii) $\Phi \subset M(T)$ the roots

iii) $\forall \alpha \in \Phi$, $\lambda_\alpha : \mathbb{G}_{a,S} \hookrightarrow G$ the corresponding root space

iv) $W = \mathrm{Norm}(T)/T$: an étale split finite group scheme over S that we can therefore regard as an ordinary finite group

v) $W_a \cong$ K-rat. pts. of Norm(T)/R-rat. pts. of T.

= group of automorphisms of $N_{\mathbb{R}}(T)$ given by W and translations with respect to N(T) = "the affine Weyl group".

vi) $\{\rho_k\}$ = polyhedral decomposition of $N_{\mathbb{R}}(T)$ defined by the hyperplanes $\left\{x \in N_{\mathbb{R}}(T) \,\middle|\, \langle\alpha,x\rangle = n\right\}$, for all $\alpha \in \Phi$, $n \in \mathbb{Z}$. It is known that W_a acts simply transitively on the top dimensional ρ_k, which are called the affine Weyl chambers.

vii) Let $\Phi = \Phi^+ \cup \Phi^-$ be a decomposition of Φ into positive and negative roots; then one such chamber is given by:

$$\rho_o = \left\{x \in N_{\mathbb{R}}(T) \,\middle|\, 0 \leq \langle\alpha,x\rangle \leq 1,\ \alpha \in \Phi^+\right\}.$$

viii) For any R-algebra R', we let G(R') denote the set of R'-valued points of G.

ix) With respect to T and Φ^+, define the <u>Iwahori subgroup</u> $\mathfrak{B} \subset G(R)$ as either

a) the inverse image of B(k) in $G(R) \longrightarrow G(k)$ where $B \subset G$ is the Borel subgroup associated to Φ^+

or b) the subgroup generated by T(R), $\lambda_\alpha(R)$ for all $\alpha \in \Phi^+$, $\lambda_\alpha(m)$ for all $\alpha \in \Phi^-$ (here $m \subset R$ is the maximal ideal).

x) Via $\mathfrak{B}$, we get the Bruhat decomposition of G(K):

$$\mathfrak{B}\backslash G(K)/\mathfrak{B} \cong W_a$$

or all double cosets represented by an element of Norm(T)(K), unique up to an element of T(R).

Now the orbit space G/T will exist and be an affine scheme, and G itself, via the canonical:

$$\pi: G \longrightarrow G/T$$

will be a locally trivial principle T-bundle over G/T. Define:

$$\sigma \subset N_{\mathbb{R}}(T) \times \mathbb{R}$$
$$\sigma = \text{cone over } \rho_0 \times (1)$$

giving us an affine embedding $T \subset \mathbf{T}_\sigma$, (note here: T and not just T_η is in T_σ because $(0)\times\mathbb{R}_+$ is a face of σ!) and hence the associated fibre bundle:

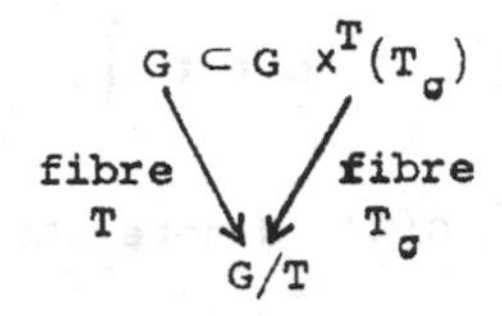

Note that $T_\eta = (T_\sigma)_\eta$, hence $G_\eta = (G \times^T T_\sigma)_\eta$. Clearly G acts on the left on $G \times^T T_\sigma$. More important is:

<u>Lemma</u>: For all $x \in \mathfrak{B}$, right translation R_x by x extends to an automorphism of $G \times^T T_\rho$.

<u>Proof</u>: We shall check that for all discrete valuation rings R' dominating R, with quotient field K', R_x restricts as follows:

$$(*)\qquad \begin{array}{ccc} G(K') & \xrightarrow{\ R_x\ } & G(K') \\ \| & & \| \\ G\times^T T_\sigma(K') & & G\times^T T_\sigma(K') \\ \cup & & \cup \\ G\times^T T_\sigma(R') & \xrightarrow[\text{res}]{} & G\times^T T_\sigma(R') \end{array}$$

It follows then from the valuative criterion that the graph $\Gamma \subset (G\times^T T_\sigma)\times_S (G\times^T T_\sigma)$ of the rational map R_x is proper over $G\times^T T_\sigma$. But since $G\times^T T_\sigma$ is affine, this implies Γ is finite over $G\times^T T_\sigma$. Since $G\times^T T_\sigma$ is smooth over T_σ, which is normal, $G\times^T T_\sigma$ is normal, hence by Zariski's Main Theorem, $\Gamma \longrightarrow G\times^T T_\sigma$ is an isomorphism and R_x extends.

Next, we need only check (*) for $x \in T(R)$; $x \in \lambda_\alpha(R)$, $\alpha \in \Phi^+$; or $x \in \lambda_\alpha(m)$, $\alpha \in \Phi^-$. Since T acts on T_ρ on the right, the first of these 3 cases is trivial. Now also, $G \longrightarrow G/T$ locally has sections, so for all local rings $\mathfrak{O}$, if we have an $\mathfrak{O}$-valued point of $G\times^T T_\sigma$, we can:

a) project it to G/T, b) lift it to G, c) hence decompose the original point into an $\mathfrak{O}$-valued point of G times an $\mathfrak{O}$-valued point of T_σ:

$$G \times^T T_\sigma(\mathfrak{O}) = G(\mathfrak{O})\cdot T_\sigma(\mathfrak{O}).$$

We are reduced to checking:

i) $T_\sigma(R')\cdot\lambda_\alpha(R) \subseteq \lambda_\alpha(R')\cdot T_\sigma(R')$, $\forall \alpha \in \Phi^+$

ii) $T_\sigma(R')\cdot\lambda_\alpha(m) \subseteq \lambda_\alpha(R')\cdot T_\sigma(R')$, $\forall \alpha \in \Phi^-$.

But by definition of root spaces:

$$t\cdot\lambda_\alpha(x) = \lambda_\alpha(\chi^\alpha(t)\cdot x)\cdot t, \ \forall t \in T(K'),\ x \in K'.$$

Let v' denote the valuation on R', extending that on R and normalized so that $v'(\pi) = 1$, π = generator of m. There is a natural injection

$$T(K')/T(R') \xrightarrow[\text{ord}]{} N_{\mathbb{R}}(T)$$

defined by

$$v(\mathfrak{X}^{\alpha}(t)) = \langle\alpha, \text{ord } t\rangle$$

and $T_\sigma(R')$ is just $\text{ord}^{-1}(\sigma)$. Hence if $t \in T_\sigma(R')$, then $0 \leq \langle\alpha, \text{ord } t\rangle \leq 1$, all $\alpha \in \Phi^+$, hence $\mathfrak{X}^{\alpha}(t)$, $^{\pi}/\mathfrak{X}^{\alpha}(t) \in R'$, all $\alpha \in \Phi^+$; hence $\mathfrak{m}\cdot\mathfrak{X}^{\alpha}(t) \subseteq R'$, all $\alpha \in \Phi^-$. Putting these together proves (i) and (ii). QED

Now, in a somewhat informal notation, we make our key definition as follows:

$$\overline{G} = \bigcup_{x \in G(K)} (G \times^T T_\sigma)\cdot x .$$

What this means is: for every such x, we take a copy $(G\times^T T_\sigma)\cdot x$ of $G\times^T T_\sigma$ and we identify G_η with its generic fibre by:

$$G_\eta \xrightarrow[\approx]{R_x^{-1}} G_\eta \subset G \times^T T_\sigma \cong (G \times^T T_\sigma)\cdot x .$$

In other words, for every $x \in G(K)$, we have open embeddings

$$G_\eta \longrightarrow Z(x) \quad \text{(generic fibres equal).}$$

In such a situation there is at most one way to glue together these $Z(x)$'s into a separated scheme Z with generic fibre G_η. To check now that our $\overline{G}$ does exist, take any 2 elements $x_1, x_2 \in G(K)$. Using the Bruhat decomposition (x) above, write

$$x_1 x_2^{-1} = b_1 t\cdot w\cdot b_2^{-1}, \quad t \in T(K),\ w \in W.$$

Then we asked to check the possibility of glueing:

$$\begin{array}{l} \quad (((G \times^T T_\sigma)b_1)tw)y \\ \cup \\ G_\eta \\ \cap \\ \quad ((G \times^T T_\sigma)b_2)y \end{array}$$

where $y = b_2^{-1}x_2$. As y merely adds an automorphism to the situation, we may drop y. Furthermore, by the lemma, for all b $\in \mathfrak{B}$, there is a commutative diagram:

$$\begin{array}{ccc} G_\eta & \subset & G \times^T T_\sigma \\ \| & & \downarrow \wr \\ G_\eta & \subset & (G \times^T T_\sigma)\cdot b \end{array}$$

i.e., these embeddings are completely identified by the closure of the graph of the identity map from G_η to G_η in $(G\times^T T_\sigma)\times_S(G\times^T T_\sigma\cdot b)$. We are reduced, therefore, to checking the glueability for:

$$\begin{array}{l} \quad (G \times^T T_\sigma)\cdot tw \\ \cup \\ G_\eta \\ \cap \\ \quad G \times^T T_\sigma \quad . \end{array}$$

But take the entire decomposition of $N_{\mathbb{R}}(T)$ into affine Weyl chambers $\{\rho_k\}$, and let σ_k be the cone over $\rho_k\times(1)$ in $N_{\mathbb{R}}(T)\times\mathbb{R}$. Let the decomposition $\{\sigma_k\}$ define $T \subset \overline{T}$: here $\overline{T}$ is a scheme locally of finite type over S gotten as an infinite union of T_{σ_k}'s, but still with generic fibre $\overline{T}_\eta = T_\eta$. Note that since $\{\rho_k\}$ is stable under W and N(T), both translations by $t \in T(K)$ and conjugations by $w \in W$ extend to automorphisms of $\overline{T}$. Now form $G \times^T \overline{T}$. Then clearly we get a commutative diagram:

$$\begin{array}{ccccc} & & (G \times^T T_\sigma)\cdot tw \cong G \times^T \left(T_{tw(\sigma)}\right) & & \\ & \cup & & \subset & \\ G_\eta & & & & G \times^T \overline{T} \\ & \subset & & \cup & \\ & & G \times^T T_\sigma & & \end{array}$$

so glueability is checked.

This completes the construction of $\overline{G}$. We summarize its properties as follows, leaving the verifications to the reader:

a) $G \subset \overline{G}$ is a toroidal embedding without self-intersection and $G_\eta = \overline{G}_\eta$,

b) the left action of G on itself extends to an action of G on $\overline{G}$,

c) for all $x \in G(K)$, the right action R_x extends to an automorphism of $\overline{G}$

d) Via this right action, the stabilizers of the strata of $\overline{G}-G$ are precisely the parahoric subgroups of G(K), and this sets up a bijection between strata and parahoric subgroups which extends to an isomorphism of the graph of the embedding $G \subset \overline{G}$ with the Bruhat-Tits building for G over R.

e) It seems to me quite likely that if Z = center of G, then $Z\backslash\overline{G}$ is regular, and that the components of its closed fibre $(Z\backslash\overline{G})_o$ which occur with multiplicity 1 are precisely those corresponding to "special vertices" of the decomposition $\{\rho_k\}$.

I would say, roughly, that $\overline{G}$ is something like a Néron model for G over R, with certain "corners" added so that its closed fibre $\overline{G}_o$ is a union of complete varieties over k.